Prais
DOMINATE
REAL ESTATE
and JAMES TYLER

"There is a ton of actionable advice in this book. James has developed fantastic formulas for surviving bad real estate markets and creating everlasting real estate success."
— Karen Briscoe, author "Real Estate Success in 5 Minutes a Day" and host of the 5 Minute Success podcast

"James has brought his fundamental perspectives about marketing and sales, and compiled it into a master plan to help professionals take their business to the next level."
— Joe Homs, Real Estate Investor, H&M Realty Group

"The concepts inside the Dominate Real Estate book are priceless. James provides a mapped out plan for any real estate agent to follow, whether experienced or just starting in the industry."
— Debbie West, Broker Associate, West Coast Realty Group

"James delivers systematic strategies for building successful real estate businesses. He showcases several techniques that demonstrate how real estate professionals can drive positive changes in their business. Achieve superior results by leveraging DRE's tactics and methods. If each of us embodies the teachings that are in this book, it would not only create succeeding outcomes for

our businesses, but we will also live better lives!"
— Salee Zawerbek, REALTOR®, Keller Williams Newport Estates

"I met James at an Orange County Association of REALTORS® (OCAR) event months before he finished his book. At that time, it was apparent that his book was a must-read. Now and after reading the first few chapters, I knew this book would complement my background in the real estate market research. James brings new practical strategies and creative methods in growing any real estate business. His book kept me interested in wanting to read more and implement his tactics immediately."
— Ruth Bruno, REALTOR®, Regency Real Estate Brokers

"In his new book, Dominate Real Estate, James Tyler provides tools to help real estate professionals understand how to leverage the ever-changing world of technology. He contributes to the fulfillment of streamlining the business digitally. He has experimented with several online platforms, and he gives us the tried-and-true ones he recommends."
— Nancy Teixeira Jenson, REALTOR®, Coldwell Banker Residential Brokerage

"In a world with an increasing amount of information with often less pertinence, James Tyler has written a primer on taking our daily activities from superficial transactions to meaningful ones. Dominate Real Estate presents sales and marketing tools, examples, and tactics to enhance any part of our professional and personal lives."
— Craig Preston, REALTOR®, First Team Real Estate

"A few chapters into Dominate Real Estate, I knew this would help me take a giant step to the top. I've already started looking at my business from a growth mindset and have been implementing the material. I extend many thanks to James for sharing his vision with us."
— Monica Martinez, REALTOR®, Century 21

"If you want to succeed at the highest level possible in your real estate profession and personal life, I highly recommend reading Dominate Real Estate. You will learn everything from envisioning your goals, setting up your entire brand, to executing lead generation tactics in the most structured way possible. Thank you, James!"
— Laura Rodriguez, REALTOR®, Signature Real Estate

"After reading the first five chapters of Dominate Real Estate, I knew right away that I would approach my Real Estate business differently. I have read many different books on Real Estate and how to command the market, but this book offers real strategies and real solution. It provides insights on how to dominate in communities. This book has me excited and motivated to Re-Launch my RE career and utilize some different techniques to become a top producer. Thank you, Coach James Tyler."
— Shawana Bowens, REALTOR®, Agnelli Real Estate

"Dominate Real Estate is an insider guide to build an entire real estate business. James delves into every aspect of business development, marketing, and sales with practical methods to survive the real estate market at any given time. Use this book as your roadmap to forge lasting success. I wish this book would have been available to me when I started in real estate."
— Keith Brodsky, REALTOR®, RE/MAX
Over 2000 properties sold in 3 states

"This book is a total game changer! Dominate Real Estate is a great find for everyone who wants to learn how to build high-quality relationships with the several networks that surround us. I have already started to recognize the importance of strengthening my relationships with people in my database and my social media to help scale my business."
— Christian Plowden, REALTOR®, Keller Williams Realty

"The vision of this book is impressive. It is a blueprint that helps take out the guessing game of how to succeed in your real estate business. James provides the real estate agent everything they need from real estate scripts, lead generation strategies, and listing presentations."
— Christopher Daish, REALTOR®, Corcoran Real Estate

"I can see the great amount of experience James Tyler has shared in this book about real estate marketing and sales tactics. I love how he gives robust strategies that are supported by research and through his wisdom over the years. I would recommend this amazing book to every real estate professional. I think everyone in the sales world can benefit from it and learn to dominate their market."
— Ricardo Leveron, REALTOR®, Keller Williams Realty

"This book is the map for the road to success. James teaches step-by-step how to

get on that road and stay on it. Dominate Real Estate will help any real estate
agent outmaneuver the majority of his or her competitors."
— Gloria Resendez, REALTOR®, Encore Fine Properties

"This book is a must-read for anyone who wants to have a significant impact on leadership. It focuses on the growth mindset and how it can influence success tremendously. Leadership is not a programmed trait, it is a hard-won skill, and learning how to lead requires a master guide. James elaborates on goal-setting, so we don't set limitations for ourselves and consistently feel accomplished. James Tyler broke the mold with Dominate Real Estate."
— Brandon Grass, Real Estate Agent, RE/MAX Kelowna

DOMINATE REAL ESTATE

DOMINATE REAL ESTATE

*A Master Plan to Build a
Thriving Real Estate Business*

ACTIONABLE SALES & MARKETING STRATEGIES
FOR REAL ESTATE PROFESSIONALS

JAMES TYLER

MARKETING
ENGINES.
STRATEGY & GROWTH

An imprint of **Marketing Engines, Inc.**

MARKETING
ENGINES.

Published in the United States by
Marketing Engines®, Inc.
27702 Crown Valley Pkwy. Suite D4-217
Ladera Ranch, CA 92694
(800) 949-3644
www.MarketingEngines.com

Special discounts are available on quantity purchases by corporations, real estate associations, educators, and others for sales promotions or corporate use. For details, contact the publisher at the above-listed information.

U.S. trade bookstores and wholesalers: Please contact the publisher at the above-listed information.

Library of Congress Cataloging-in-Publication Data has been applied for.
Name: James Tyler, the author.
Title: Dominate Real Estate
Description: First Edition | California: Marketing Engines, 2018.
Identifiers: First Paperback Edition ISBN 978-1-7335033-0-3
Subjects: LCSH: Success in business. | Consumer behavior. | Marketing
BISAC: SELF-HELP / Personal Growth / Success. | BUSINESS &
ECONOMICS / Consumer Behavior. | BUSINESS & ECONOMICS /
Marketing / Direct. | BUSINESS & ECONOMICS / Leadership. | BUSINESS
& ECONOMICS / Development / Business Development. | BUSINESS
& ECONOMICS / Investments & Securities / Real Estate. | BUSINESS
& ECONOMICS / Real Estate / Buying & Selling Homes. | BUSINESS &
ECONOMICS / E-Commerce / Internet Marketing.

ISBN 978-1-7335033-0-3
Audiobook ISBN 978-1-7335033-1-0
International edition ISBN 978-1-7335033-2-7

Printed in the United States of America

Book design and illustrations by James Tyler

DEDICATION

*F*ind out what your purpose is in life, and chase after it with passion, knowledge, and everything you have — dedicated with love to my family, my stunning wife Randa, and our precious daughter Liana. I will always love♥you more than you will ever know.

CONTENTS

CONTENTS

ACKNOWLEDGMENTS

I am humbly thankful, forever grateful, and honored that you have picked up this book. There is nothing more satisfying than feeling a great joy within. So the least I can do is to say thank you for bringing joy to my heart and reading my book. I hope you find words of wisdom, practical value in what I have to share, and satisfaction in the prosperous outcomes that will come into your life.

I want to thank my wife, my family, and my friends, for their patience and continued support. In so many ways, I couldn't have written this book without them. A special thanks to Bshara Salem, author of "Marriage: The Earthly Heaven" for his enlightening wisdom and mentorship. Also, I want to extend a notable appreciation to the people that had a positive influence on my life. I firmly believe each one has contributed to what I know today.

There isn't space to individually thank all the people who have helped me learn the lessons that were the spark of this book. Also, those who have influenced me in ways I couldn't have imagined. Notwithstanding, Theodore Levitt, Zig Ziglar, Todd Duncan, Ron Willingham, Anthony Robbins, and Sean Ellis deserve a special mention.

Many real estate agents invested their time to help me shape my ideas and speculations into coherent topics. Other agents provided vital feedback, background, and insight. To this end, I would particularly like to thank all of the real estate professionals who contributed and supported in one way or another—whether offline, online, or social media. To name a few, I extend my appreciation to Adam Alcaraz, Deborah West, Jessica Tomelden, Joe Homs, Michael Darwich, Arne DeWitt, Gloria Resendez, Brandon Grass, Nancy Jenson, Liz Deering, Craig Preston, Keith Brodsky, Rosaura Santos, Laura Rodriguez, Christopher Daish, Akbar Mubarak, Renato Torres, Liz Ramos, Stephanie Cure, Raul Briseno, Adam Stinson, Salee Zawerbek, Kristen Hoerner Alrayes, Ricardo Leveron, Frank Benevento, Terri Day, Luisa Ayala, LaTrice Davis, Aolany Gonzalez, Kevin Joelson, Julian Chavez, and the many others that I am unable to mention all of their names.

Last but foremost, I want to extend a significant appreciation to my wife, Randa, for being patient through the evenings and weekends I was locked away in our office writing this book. Also, for taking care of our newborn daughter, Liana, while creating this excellent master plan for you to thrive.

Other Acknowledgements

I am just grateful to the world and everyone in my life that participated in making me the person of which I have come to be today!

Thank you!

FOREWORD

James Tyler's desire to help real estate professionals nationwide reach the highest success measure in their local markets is apparent in everything he does and is reassured this remarkable book.

James has acquired tremendous and invaluable business knowledge over the last two decades. He reflects his real estate, marketing, and technical expertise in this masterpiece. It is evident to me that James is an innovative leader with a creative mind. As you read the pages, you and your company will benefit from his marketing, sales, and technical intelligence.

This book mirrors the principals every company needs to thrive. It will empower you with the fundamental knowledge necessary to become a great leader and a top producer in the real estate industry. Regardless of your current experience, this is a must-read and must-follow guide for everyone.

The methods, tactics, and all of the information this publication contains will form an unstoppable momentum in your career. You're holding in your hands a real domination system. I genuinely believe it will become one of the cornerstone books for your business and the real estate industry. Get compelled to maximize your potential in real estate and your life.

Randa Wahid, M.H.A.
Clinical Project Manager II

PREFACE

Magnus Carlsen, a chess grandmaster and the current World Chess Champion—his peak rating of 2882, achieved in 2014, is the highest in history. He uses a variety of openings to make it more difficult for opponents to prepare against him. "Checkmate" is the goal of the game.

The game of chess requires control over a complex set of skills that are both an art and a science. Players must be attentive and aware of the strengths and weaknesses of their situations, and those of the opponent. Players have their strategies; some are short term, and others are long term. While some moves might be predictable, and others are influential, each movement has an impact and is also a step forward in positioning for a winning endgame. Hence, flexibility within each strategy is critical to achieving the goal, which is to win the game.

Like chess, the real estate playing field requires an artful blend of knowledge and a scientific mix of tactics. However, while a chessboard is limited to 64 squares and is two-dimensional, the real estate industry requires thinking across multiple dimensions from economic and political, to social and technological.

Novices may often spread themselves quickly and aggressively into the fray, seeking swift advantage but overlooking the effect of countermeasures that are obvious to knowledgeable, skilled, and experienced players. Tactical thinkers see beyond the "next move." They anticipate the progress of a series of movements. Combining these steps and actions will create a dominant strategy and control of the board.

To experience high enthusiasm, and win with passion and confidence, you need to "Checkmate" the top players in the space. A real estate professional must look beyond a single move, understand multiple marketing and sales key concepts, and have the knowledge and the skills necessary to execute a carefully plotted plan. This plan consists of concise strategies all linked to form an evolving, powerful program geared towards one goal; *Dominate Real Estate*.

I have designed this book to provide you with the knowledge and the tools that will make you more effective in your real estate career and activities. Get ready to embark on the essential Business Development, Marketing, and Sales concepts that will be brought to life as you go through the chapters. Apply these methods and tactics to your particular company and situation. By doing so, and while reading this guide, not only you will be growing your business but also building a reliable Domination System.

Unfortunately, for many small, medium, and sometimes large companies, the concepts in this book are either not understood, or merely unknown. By believing and applying them, you will be at an enormous advantage over a vast majority of your real estate marketplace.

— James Tyler

IMPORTANT:

Access the "Dominate Real Estate Methods & Tactics" at https://Dominate.RealEstate

Take the Next Step
Execute Tactics 0.0 and 0.1

INTRODUCTION

There are millions of real estate professionals nationwide competing for the same buyers and sellers. All of them, in one way or another, are striving to be successful. The good news is that our population is growing at a high rate, and the demand for more real estate is growing with it.

On the other hand, competition is extremely high; technology is advancing at a soaring rate that we can't begin to fathom. Staying competitive and keeping up with the vast majority of high tech has become very challenging, complicated, and somewhat cumbersome. Where is the future of the real estate industry headed? How do you prepare yourself for the unknown? How do you build a reliable system that will guarantee your success?

Dominate Real Estate

Bearing in mind what I have covered in the Preface of this book. The merit of the chess story was that if you want to dominate and take over your marketplace, you have only one choice. You must become the number one producing real estate advisor in your territory. While it is a great honor to rank second, third, or whatever level humbly satisfies you—your primary goal is to rank first. It would be best if you become the number one real estate expert and the agent of choice in your marketplace at all times. Not to feed your ego, not to be the best salesperson, not for the money, but to accomplish and earn anything to which you put your mind. By being the most outstanding individual at serving your clients.

Top real estate producers have figured out one way or another to keep producing. Here's a little secret: **Build a working system, then a profitable brand**. For every strategy you pursue, you must start small, provide value, hustle, offer more value, test, improve, give even more benefit, scale, and then dominate. For example, your goal should not be "I want to become the number one real estate agent in a given city, county, or state." A wise, experienced agent would refine this goal to become the number one in the surrounding 500 homes only in a specific community or neighborhood. Once top agents understand their community needs and can serve their wants with a working system, only then will they see positive results. They will then expand their reach to help a bigger audience.

Effectively engaging your neighborhood is the foundation for success, growth, and dominance. Initially, try to become the number one agent of choice for a small community. For example, choose one hundred homes in your neighborhood and measure how long, and with how many people you can build meaningful relationships. When it is time to buy or sell, you should be the real estate agent they'll undoubtedly choose. Imagine setting a goal to dominate more significant communities higher than ten thousand homes.

Real estate professionals must grow their customer base to survive and thrive. However, taking over the market isn't just about how to get new customers. More importantly, it's about how to activate, engage, and build relationships. It is about winning far more people in the quickest time possible, so they keep coming back. It is about elevating your values and your brand. It all starts by building a system that will continuously create more opportunities. More deals will turn into loyal customers; then passionate ambassadors will ultimately become your reliable referral engines.

Furthermore, the lack of vision clouds the success system and your mission to become the agent of choice. Imagine for a moment that you have a magic wand. Wave it to portray the perfect real estate business that is guaranteed success until eternity. How does that place look like to you? What would you do to get there? That is the pinnacle of success! Seeing where you are right now and where you want to be will clarify the journey that will get you there. Do you see all of the steps? How far, how long, how high, how easy, and how tough? Do you have the knowledge, the patience, the endurance, the dedication, the commitment, and the energy to get there?

In this book, and together, we will pioneer the essential methods that will evolve and interconnect with one another to form a reliable growth system. A dynamic system that will become your business model; with it, you will attract, nurture, convert, monetize, and dominate your real estate market. Together we will create the freedom you've been seeking for many years. Live a better life, spend more quality time with the family, travel the world, or whatever your heart wishes for—without frustrations or large budgets, but with hard work and dedication. Yes, you have to work to succeed, grow, and dominate.

Before diving in, let's take a look at where the real estate industry currently is and where it is heading.

Shift Your Mindset from 'Human vs. Machine' to 'Human plus Machine'

There are many speculations about Artificial Intelligence taking over humanity because robots are going to outsmart us in many different ways and many aspects of our lives. I am neither here to predict the future, nor am I here to judge machines. I am here to advise you that technology isn't your enemy; it is your ally, embrace it. If you are not tech-savvy, and technology isn't your best friend, you need to shift your mindset from Human versus Machine to Human plus Machine. Take advantage of technology before you miss the bus.

Consumer buying and selling cultures have changed quickly and tremendously from offline to online, and mobile. You are currently witnessing how social networks have been playing a significant role in people's lives. Social platforms determine how people behave online, whom they associate themselves with, where they live, want to live, work, and recreate.

As you may have already noticed, we are losing the human touch in many aspects of our lives. The real estate industry is no exception. How difficult has it been for you to get people on the phone? How about getting with sellers face-to-face? What about making an offer in person while your buyers and sellers are present? These types of behaviors seem so passé and outdated. The reality is that they are still the most robust sales tactics in converting clients.

While customers' culture and online behaviors have shifted to digital, some aspects of the real estate business are more effective if kept old-fashioned. Machines do well with low-level processing tasks, which you can use to improve your work and save time. It is essential to recognize that this is still a partnership. The machine is optimizing from a set of variations and can automate specific tasks. However, robots cannot write an ad from scratch, communicate with a potential client on a deep level, and understand the client's needs or their emotions—at least not yet. After all, you will discover what tasks can be automated by machines, and what tactics must be driven by real humans. Most importantly, this book will help you shift your mindset from Humans Versus Machines to Humans plus Machines. You will take advantage of the most advanced technologies to scale your real estate business rapidly and effectively.

> **IMPORTANT:**
> *Access the "Dominate Real Estate Methods & Tactics" at*
> *https://Dominate.RealEstate*
>
> **Take the Next Step**
> *Execute "Tactic 0.2: How to Become Agile and Measure Instantly!"*

The Future of Real Estate

Scoring more listings and qualified home buyers seem to be very challenging tasks for most real estate professionals. How can you do both effectively now and in the future—by building a flexible system that is repeatable, scalable, and predictable.

Be Agile: Technology is moving industries and customer behaviors fast. Generation Y moved from Facebook to Snapchat and Instagram in less than five years. You need to master all social channels and be as agile and flexible as your audience. When they move, you move.

Be Transparent: Despite the apparent flux of technologies and the flexibility of the real estate industry to pivot—you must not lose your core existence. Your ability to service your customer needs supersedes any trending technology. Marketing and sales in the upcoming years will be more about engagement, satisfaction, and relationships. The more transparent and available you are to your customers, the more effective your business will be. Hence, there will be less focus on Customer Relationship Management Systems (CRMs) and more emphasis on Customer Engagement Hubs (CEHs).

Measure Metrics and Analytics: Big data has created a massive shift in marketing in the past few years, affecting all industries, including real estate. Data scientists and marketers are facing many challenges making sense of the data. Using historical data, and then going through a lengthy hygiene process, data becomes outdated due to the constant flow of new data into organizations. For example, by the time data scientists clean ten years' worth of information to decipher the good from the bad; new data emerges, making historical results obsolete. Hence, don't try to measure every metric in your business. Focus on the metrics that matter, measure quickly and often. If you want to grow your business, metrics and analytics should be looked at daily.

Build Influential Networks: With traditional marketing, we see celebrities advertise on TV. Now, and in the next ten years, digital will become far more prominent than TV, driving actors, actresses, and celebrities to advertise on websites, social channels, mobile, and IoT devices. The digital world is saturated with celebrities and top influencers. They are significant drivers in marketing and increased sales. Generate more business by building vendor relationships and partnering with top influencers in the real estate industry. Lead generation will heavily focus on online referrals from digital partners than consumers.

Tell Creative Narratives: Effective brands are those that have good stories. A story is a powerful tactic to attract attention. It has a beginning, middle, and an end that creates an emotional connection and engagement. However, brands will be moving into telling a "creative narrative." Instead of telling a story about your brand, to thrive, brands need to tell stories about their customers. A narrative is open-ended, it is about an opportunity, and whether the listener gets the benefit depends on the listener's choices. The resolution has not yet occurred. You're talking about some opportunity that hasn't yet materialized and the ability to embrace this opportunity hinges on the listener's actions. A narrative is a call to action. It says, "How it ends is up to you. What are you going to do?" Storytelling and creative narratives will focus on inviting people to brand instead of interpreting through advertising. What do you have or what does your brand do that's worth all the hype?

Predict, Personalize, and Automate: Google, Apple, Waze, and many Fortune 500 companies leverage data and predictive analytics to predict potential customer needs and behaviors. This way, brands can personalize their messages and automate conversions. Personalization and automation using Big Data and Artificial Intelligence (AI) will be utilized in every technology moving forward. For example, notice auto-completion while entering a destination you've already visited in Waze or writing a sentence in Google Mail (Gmail). These applications can predict what you are about to type automating solutions so you—the customer—aren't inconvenienced with manual tasks saving you time and hassle. Learn how to take advantage of AI-based technologies for your business. Use Predictive Analytics to personalize your messages, content, and services to your customers so you can earn more of them.

Think IoT for Business: You still go to the office, sit on your desk, and work on your desktop or laptop. In the next decade, you will no longer need an office desk. Your business will become strictly mobile. The rise of mobile usability and usefulness has evolved but not to the extent where we're 100% dependent on the Internet of Things (IoT). You will scan documents, print files, email campaigns, text in mass, e-sign, process payroll, and perhaps file your taxes—not just with your phone— but with wearable devices such as Apple Watch, connected clothes, Bluetooth glasses, 3D goggles, and more.

AR, VR, AI, IoT, and More

With the proliferation of Big Data, Artificial Intelligence, Predictive Analytics, Internet of Things, and Mixed Reality (a combination of Augmented Reality and Virtual Reality), advanced technologies will be transforming the real estate sector even more in the next few years. As of this writing in 2019, technology hasn't fully evolved in the real estate industry yet. There is undoubtedly more to emerge.

Consumers will make far more use of mobility and digital assistant technologies that facilitate behaviors, processes, and telecommunication. The method and popularity of virtual Open House tours will increase. Mobile applications and robots fulfilling basic real estate needs will be more prominent in the future. Most agents are overwhelmed with the proliferation of technology, websites, and mobile apps. Nonetheless, far more online platforms and mobile applications will appear. Most of them will sell you, the real estate professional, on better quality leads, AI-powered leads, personalized marketing automation, and enhanced sales predictions. It is going to be much harder for you to ignore the hype and filter through the noise. So, how do you reinvent your real estate business, and avoid becoming obsolete while staying competitive?

Real estate has been and will always be a personal-driven business. It is a relationship-based and high-trust type of industry. Real estate professionals must fear none. Embrace the chaos and learn how to leverage data and technology to your advantage. Cultivate far more people, and build better relationships to change the outcome significantly. Begin by really understanding how to work with the available data and the latest technologies to swiftly excel at what matters—building meaningful relationships. Data and technologies covered in this book will change your perspective positively on how you live and how you'll run your real estate business.

Here's an example of where the future of real estate is moving:

Okay Google Voice Searches & Actions

Buyer: "Okay, Google. Is there an Open House around my area this weekend?"

Google Assistant: "Great question. What time are you looking to visit the Open House?"

Buyer: "Between 3 and 4 pm."

Google Assistant: "How many bedrooms and bathrooms are you looking for?"

Buyer: I am looking for four beds and three baths.

Google Assistant: Okay, there are two Open House events in your area that fit your needs. There is one on Main Street, and another on 7th Street. Would you like to see those two homes?"

Buyer: "Yes. I will swing by there this afternoon."

Google Assistant: "Great. I will put them on your To-Do list and add a reminder on your calendar."

Moving forward, it is highly imperative that you integrate every aspect of your real estate business with available technologies. Broaden your reach by allowing smart devices, online platforms, and advanced technologies to seemingly communicate with your brand leading potential clients straight to your doorstep.

iBuying Real Estate

Your most significant competitors are trying to put real estate brokerages out of business. If you haven't heard about iBuying, it is now time to learn it. Soon, most sellers will challenge you on why they should use you instead of iBuyers. You need to know how to overcome such an objection.

OpenDoor, PurpleBricks, OfferPad, and similar platforms are on the rise. Venture Capitalists (VCs) are investing billions of dollars in these internet home buying giants. Their goal is to revolutionize real estate buying online without the need for the middle man—the real estate professional.

Overcome this challenge by advising clients that these companies don't have "Fiduciary Obligations" or "Full Disclosures." They don't have Real Estate Sales Agreements which protects the customers legally and federally. Real estate agents look for their clients' best interest. Buyers and sellers get federally protected under the Community Reinvestment Act (CRA) and Housing Urban Development (HUD).

Utilizing iBuying companies to buy and sell real estate, thereby waiving all your fiduciary rights as a customer. In other words, you would be selling and buying homes online, "As-Is." These iBuying companies look for their own best interest, not yours as a customer, nor mine as an agent. Their "As Is" Disclosure Agreements protect them in case of disasters. That is why it behooves people to work with federally licensed real estate professionals who would make the best recommendations instead of these companies:

- **OpenDoor**: https://opendoor.com
- **PurpleBricks**: https://purplebricks.com
- **OfferPad**: https://offerpad.com
- **Knock**: https://knock.com
- **Zillow Instant Offers**: https://zillow.com/offers
- **Redfin Now**: https://redfin.com/now

DISCLOSURE

I am not an attorney, nor am I providing legal advice. Please check with your local real estate attorney for information about iBuying laws.

How Is This Book Different

You have put your trust in me as your coach by purchasing my book; therefore, I will not waste your time. I know that you're busy, and I understand you have important things to do with your life; hence, I want to take a quick moment and thank you for trusting in me. I feel honored that you've decided to spend your valuable time with me. I also wanted to tell you that this book is different in many ways.

This Book Is Evergreen

Fast forward a decade, most of the information in this book will still apply not only to you and your real estate business but to most companies in many industries. I am not only sharing with you the strategies and tactics that will work in the current market, but I am also sharing the principles of those strategies and their foundations. In case there is a market shift—and there will be—my theories won't change. You'll be able to pivot accordingly just as the Chinese proverb suggests:

"Give a man a fish, and you feed him for a day; teach a man to fish, and you feed him for a lifetime."
—Pin Yin

I Lead by Example

You need to know that the information in this book is trustworthy and proven to work. This book is a mix of over two decades of experience I acquired in leadership, technology, marketing, sales, and finance—specifically real estate. I have coded and designed web applications since 1994. Since the early 2000s, I have worked in every possible mortgage and real estate position. I started a marketing agency in 2013, specializing in the financial industry. Just recently, I started building a real estate platform powered by machine learning and artificial intelligence. Rest assured the information you're about to embark on is genuine and authentic. Please don't take anything lightly no matter how simple or complex it may sound to you. Get ready for a wealth of information that is going to transform not only your real estate business but also your life.

"When eating an elephant, take one bite at a time."
—Creighton Abrams

How The Book is Organized

Businesses of all sizes, in every industry and all around the world, are desperate for quality customers. Most are struggling mightily to find creative and innovative ways to grow. Domination isn't even on the radar for most organizations. In this book, I will provide rigorous methodologies for driving online traffic, creating opportunities, and converting leads to loyal customers. We will rely on data-driven analysis and experimentation. My methods will use the power of the wealth of data and technologies to create a system that will positively influence the outcome and the growth of your business.

I will uncover critical information on how to position yourself and your brand, so clients choose to work with you. Furthermore, you'll understand the fundamentals of business development, internet marketing, and sales allowing you to create and implement effective campaigns. You will rely on proven sales scripts, data-driven analysis, and practical experimentation.

In the end, you will own a robust marketing and sales growth system, outshining the competition and capturing new customers while retaining your current ones. Whether someone is buying or selling, you will be the agent of choice to satisfy your clients' wants and needs. Buckle up, and enjoy the ride to dominate your real estate marketplace.

I have divided the chapters in this book into four different parts:

1. *PART I. LEADERSHIP:* I want to teach you the foundation of leadership quickly. Here, you will learn the legacy leaders' traits that are necessary to thrive and succeed in any venture in life, not just real estate. The goal is to create a mental shift and change your belief system for trusting in yourself.
2. *PART II. MANAGEMENT:* Setting and achieving your goals begins with your vision. Learn how to leverage time, people, and money to reach your target quickly and effectively.
3. *PART III. ART:* Crafting unique value propositions and enticing brand messages is at the heart of every successful business. You will learn how to write compelling headlines— sometimes called "hooks," as well as story-lines, and enticing calls to action so you can pursue others into working with you.
4. *PART IV. SCIENCE:* Here, you will apply everything you learned from the previous three parts. We will dive into business development, marketing, and sales. They are the most essential three departments every company needs to run a thriving business. By the end of this book, you will have the foundational knowledge to execute any growth strategy with ease.

Let's summarize each chapter now that you have had a glimpse into the four parts of the book.

Chapter 1: Become a Legacy Leader

Understand the difference between the Growth Mindset and the Fixed Mindset. Shift your frame of mind into growth in every aspect of your business using the 5 C's of the Growth Mindset: Clarity, Confidence, Commitment, Challenge, and Change. Leadership is a talent acquired and perfected over time. Change your habits, your traits, and your lifestyle to match legacy leaders. Take action to make a positive difference in your business and your life.

Chapter 2: Articulate Desires into Tangible Goals

Learn how to set short-term, mid-term, and long-term goals. Develop an execution plan to manage your time and those who will help you achieve your goals. Leverage time, people, and technology to speed up sales and growth. Learn the different levels of performance and indicators to measure success. In this chapter, you will portray your vision into tangible objectives.

Chapter 3: Discover Your Core Values

You can spend hundreds of thousands of dollars into advertising and purchasing leads. Nevertheless, if your personal and business brand values do not align with your clients' needs, your investment would end up being a huge loss. In this chapter, you'll learn how to craft your niche, your value proposition, and most importantly, discovering all of the values that will elevate your uniqueness. Most mentors and coaches tell you that you need to separate yourself from the crowd to shine, but no one breaks down the fundamentals and the foundational elements of what makes up your UVP. Get ready to be wowed.

Chapter 4: Elevate the Power of You

Since real estate isn't a tangible product that you can sell, and since most real estate services are identical in their offerings, the only entity left to make the difference is "You," the agent. To strangers, the more you are transparent, the more they'd connect with you. Let me show you how you can create a compelling tagline, an elevator pitch, and a unique value proposition. Learn the characteristics of writing a good story. Let's write an excellent biography and an executive summary that shows your potential clients your professionalism and experience. You are an experienced real estate professional, right? Most agents neither have a biography nor an executive summary that outline the real benefits of working with an agent. People want to know you and trust you before they hire you. Tell them what you are made off.

Chapter 5: Speak Your Message with Influential Content

There are three main elements to generate online leads, a hook, a story, and an offer. The hook is the enticing headline that catches people's attention. The story is what keeps your audience interested, and the offer is your excellent real estate services that acquire leads with persuasive calls to action. Here you'll learn the art of writing excellent sales copy that will make your brand stand out from all others. This chapter will focus on how to create branded content to attract and convert clients. Whether via a blog article or a video, influencing your audience begins with compelling content.

Chapter 6: Sharpen Your Axe

In this chapter, you will set up your web presence, online tracking, websites, blog, IDX website, and social media platforms. You will also sign up for management and automation platforms. You will set up all of the technologies and tools necessary for your business to thrive, succeed, and dominate. This chapter is the heart of the book and your business; without it, your growth system won't work.

Chapter 7: Build Massive Momentum

Herein I will reveal the best tactic that I have been using for many years, and with several companies to build massive databases. You will learn how to create unstoppable momentum and increase your revenue as a result of mass targeting. Collecting the right data, segmenting the right audience, and farming the proper territory solidifies your foundation for better marketing and sales campaigns. While this chapter may be a bit technical, it is what will transform your real estate database and momentum.

Chapter 8: Plant More Seeds

Sync your database with all of your connected devices. Have immediate access to your business and your clients on the go. Begin your business development direct outreach to introduce yourself and your brand. Create awareness and prepare for marketing. In this chapter, you'll set the tone and inform your audience that you exist.

Chapter 9: Generate High-Quality Leads

Being omnipresent is the key to lead generation. In this chapter, I will share with you the best offline and online strategies for generating quality real estate leads. I will take you on an exciting journey from local branding to the press, then from content to social media. We will not stop there! You'll also learn about online traffic and which type is best in not only capturing clients but retaining them as well.

Chapter 10: Nurture Meaningful Relationships

Bridging leads from cold to hot is done with content and remarketing. Nurturing your existing database and cultivating relationships is crucial in converting people to clients. Here, you'll learn how lead nurturing plays a huge role in lead activation and creating sales opportunities.

Chapter 11: Create Promising Opportunities

In this chapter, you'll try to convert all nurtured leads into opportunities. Prospecting is a massive part of lead conversion as you can't create opportunities without speaking to people. Part of effective prospecting is mastering your craft. Here, you will access pre-written sales and handling objections scripts for buyers, sellers, for sale by owners, and expired listings. We will close this chapter with the importance of following up in closing sales.

Chapter 12: Seal the Deal

Sealing the deal is about closing transactions and generating more real estate sales. I explain lead monetization working with both sellers and buyers. I then dive in more depth on how to conduct an effective listing presentation, including staging, the marketing plan, and negotiation. We conclude this chapter on how to work with buyers and their financing options.

Epilogue: What Happens Next?

Uncover the three main elements to innovate, grow, and dominate any market, in any industry, and at any given time. Execute the five high-level steps to build your real estate domination system. It doesn't stop there! We will close the journey with the 10 P's to amplify your performance, become the number one real estate agent, and dominate your marketplace.

— James Tyler

IMPORTANT:

Access the "Dominate Real Estate Methods & Tactics" at https://Dominate.RealEstate

What to do next?

Pull out your Google, Android, or iPhone calendar. Schedule at least one hour a day to execute on at least two chapters a week. You should finish all twelve chapters in this book in no later than six weeks.

Grab a highlighter and a stack of sticky-notes. Whenever you run into a great thought or a creative new idea, jot it down. Highlight your information trigger, write your notes, and mark the page it's on so you can go back to it.

PART I
LEADERSHIP

CHAPTER ONE

BECOME A LEGACY LEADER

"Do what is right by you, so you can do what is right by others."
—James Tyler

Rose, the best real estate broker in the nation, is out of bed at 4 AM. While the coffee is brewing, she grabbed her phone and recorded a video, "Good morning Ross, rise and shine my love; our coffee is almost ready!" He woke up to her incoming text message. Ross had just passed his real estate exam. Despite his lack of real estate knowledge and experience, Ross feels humbly confident and super excited to enjoy his new venture with Rose.

Rose is a meticulous listing agent that loves helping sellers. She is assertive but yet unpretentious. Bob has elevated her feeling of empowerment to hold her position as the top listing agent in her county. Instead of referring potential buyers to colleagues, her husband, the selling agent, gets to represent them. Thus, they will bring in higher household revenue! Ross, of course, loves working with buyers. In reality, he has no choice. He can't step over Rose's toes. After all, she is the boss!

He wants to be the top real estate agent of choice in his local community, and Rose wants to dominate the entire real estate industry. Regardless, they must possess the traits of leaders to accomplish their goals.

Whether your vision is to become the top agent in your territory or the number one agent in the nation, you too need to establish the characteristics of legacy leaders. In this chapter, you will learn how to manifest the leader's traits necessary to incorporate into your daily experience. These qualities will make you the best at whatever you do. Once you master those attributes, you will become the thriving leader of your real estate business and your life.

Dominating begins with your first action every morning, and then every proceeding activity. Regardless if you decide to wake up to a cup of coffee, or a six-mile run; remember, it takes practice to master a craft. Some of us face a challenge to wake up in the morning, let alone think about the overwhelming daily activities. Leading is no different. It is a skill that we must develop and perfect within ourselves before we can inspire others. Take action and practice your craft so you can lead and dominate.

As leaders, we learn what to do that is right by us, so we can do what is right by others. Our path to leadership starts by being humble and modest. We progress through a systematic discipline of under-promising and over-delivering. Leaders manage the process and the results rather than the promises. What does this mean?

When making decisions, you must systematically think about the process it will take to achieve the desired effect, before jumping into the conclusion of promising merely to promise. Being considerate of other people's emotions and the fear of disappointment bring a sense of humbleness. Modestly think through the process of reaching the expected outcome and promise within your caliber. In other words, don't be enthusiastic and eager to get a listing only to have a deal in the pipeline. It would be best if you made sure potential clients are a good fit and that you can deliver on their expectations. Also, instead of finding joy in closing the sale or getting your commission check, aim at making your clients' buying or selling journey enjoyable from beginning to end. Leaders create breakthroughs by focusing on the process, not just the results.

Moreover, a leader has the quality of positively influencing people into wanting to do what is necessary. However, it is not the leader's decision to force anyone to follow them. People usually need to allow leaders to guide them. They are often self-motivated through the motives of the leader. Show your followers what drives them to be better. Inspire people to see the potential they can become, and allow them to work towards it. Therefore, as a leader, you are an inspirational guide. It's not up to you to lead others; it's up to them to follow you.

When leaders educate, inspire, and motivate others; people see greatness in those individuals and follow them. Thus, we must know the preeminence within ourselves before we can become leaders. Take the initiative and lead yourself to be successful so you can drive your business and guide everyone around you. Demonstrate great qualities in who you are, what you do, and what your purpose is in life. Leading with a meaningful cause will influence your marketplace into following you, hence, becoming an effective leader —and the real estate agent of choice.

As real estate professionals, it helps us to find our inspirer, that person who has the traits, behaviors, and process of success down to a tee. Looking back at Rose and Ross, she was his inspiration. He decided to pursue real estate after witnessing the great achievements his wife has been accomplishing. Hence, allowing him to be influenced by her success, putting Rose in the front seat of leading his real estate career.

"A Leader's Mindset is a Growth Mindset!"
—James Tyler

Your Mindset

How can we cope with the enormous choices we have to make every day? We consistently face a never-ending stream of decisions from the second we crawl out of bed in the morning. Every action we take is a result of a choice we make. Once you started reading this book, you were driven to take action, but how do you know whether the efforts that you're about to embark on are the right decisions?

When we have the right mindset, we tend to know —or at least we think we know. For most, gut feeling is a sufficient indicator for making the right move. Though it is imperative to rely on what we believe is correct, it is essential to have a logical reason behind our decision-making. There must be a mental shift and reframing of the mindset so we can uphold exceptional situations, take more significant efforts, and achieve higher outcomes. This type of performance begins by understanding the difference between *"The Growth Mindset"* and *"The Fixed Mindset."*

Rose executes her real estate strategies, learns from her mistakes, improves, pivots, and scales. Ross wants to make his wife proud. He is afraid to fail, so he tries to perfect every step he takes. In your opinion, who do you think has a Growth Mindset, and who has a Fixed Mindset? More importantly, who do you think is making smarter decisions to achieve more success, and why?

"Life is a matter of choices, and every choice you make makes you."
—John C. Maxwell

The Fixed Mindset

People with a Fixed Mindset believe that their talents are innate gifts and fixed traits. They avoid risks and failures at all costs. They mostly adhere to what they know best and never challenge themselves. In turn, they live a fixed life and never experience any change or growth. They make little effort and expect high results. It isn't their fault per se, but rather their lack of knowledge due to this Fixed Mindset and generally their unwillingness to change. Rather than learn and innovate, they lack creativity and duplicate. They are not risk-takers and prefer the safe side. In a nutshell, they are followers. You want to avoid having this type of mindset. Instead, your objective is to thrive, innovate, and possess a Growth Mindset.

For example, a real estate agent with a Fixed Mindset would read this entire book before taking one single action. For some, they might give up before getting halfway through the book or skipping through chapters to finish, without tangible progress or notable results. Individuals with a Fixed Mindset want to learn everything before taking any steps due to their fear of making mistakes or failing, and may not end up performing at all. Therefore, ironically, they fail.

People with a Fixed Mindset believe:

- They don't make any mistakes.
- When they finish whatever they're doing, it's perfect.
- When they make something perfect, that eventually becomes effortless.
- They can do something others can't do.

Fixed Mindset thinking results in:

- A refusal to take risks due to the fear of failure.
- A perception that failing is losing.
- A false sense of superiority weakened by a deep understanding of self-doubt.
- A need for validation and reassurance in themselves.
- A desire to prove oneself and show the ability to succeed.
- A theory that hard work is for the ungifted and untalented people.
- A propensity to blame outside circumstances or others when

failed.

If you're relating to this, I have great news for you. Become agile and shift your mind to *growth*.

The Growth Mindset

In contrast to the Fixed Mindset, the Growth Mindset is the attitude most influential leaders develop and improve over time. These are the visionaries who are committed to achieving great success —all the time! They are performance-based and result-oriented types of people. Due to their focus on learning and innovating, they are continuously visualizing new creative methods to fulfill numerous accomplishments. Through dedication and hard work, these leaders never stop learning, sharpening their minds, and honing their skills and talents. They acknowledge the importance of how small daily efforts can contribute towards prosperity. Known for their humility, they tend to be risk-takers who accomplish remarkable results in life.

For example, real estate professionals with a Growth Mindset reading this book would apply every principle as they move through every section and each chapter. These individuals would **execute each tactic, test the results, learn by failing, improve by doing, and dominate by practicing**.

People with the Growth Mindset are proven to achieve incredible results versus those with the Fixed Mindset. Thus, you are either a leader or a follower. Shift your mindset to growth and lead with mindfulness. Change the way you think, your daily activities and habits radically. Understand what you want and be confident in taking impactful actions daily. Regardless of the risks, you are willing to accept, stay focused on the journey and the results. Whatever massive unfolding, you are eager to uncover, never stop learning, and improving. Always look for better and faster ways to influence the outcomes. Your little actions will deliver incremental results, and your practice and efficiency will produce continuous growth.

People with a Growth Mindset believe:

- They are smart when they execute, make mistakes, and learn from them.
- When they finish something, it is acceptable.
- They can always improve something over time.
- They can do something difficult, even if never done before.
- When they work on something challenging and finally figure it

out, they are accomplished.

Growth Mindset thinking results in:

- The ability to learn from mistakes and failures.
- A passion for learning, self-improvement, and feeding the mind.
- A desire to always be challenged.
- A willingness to put in the hard work for meaningful results.
- A belief that you must measure to control the outcomes.
- A perception that practice and effort make perfect.
- Emotional resilience and ability to adapt to stressful situations.

Shift your mindset to become a modest growth leader and learn even more by following *The Five C's of the Growth Mindset*.

The 5 C's of the Growth Mindset

Here are five C's to develop a Growth Mindset:

1. **Clarity**: Know what you want out of life and execute on those goals without seeking approval from others.
2. **Confidence**: Be confident of your talents and hard work to accomplish success.
3. **Commitment**: Your tenacity and determination will overcome most risks and failures.
4. **Challenge**: Dare yourself in embracing and overcoming your weaknesses by always learning and improving.
5. **Change**: Strive to make a positive change in yourself, your life, and those around you.

> *"The more I learn; the more I learn that there is so much more to learn."*
> —James Tyler

IMPORTANT:

Access the "Dominate Real Estate Methods & Tactics" at https://Dominate.RealEstate

Take the Next Step

Execute Tactic 1.0: Commit to Leadership

Your Personality

We are all different; however, as the saying goes, "Birds of a feather flock together." We congregate with people of the same interests and qualities. Regardless, we all gravitate towards those who are confident, cheerful, and humble.

Optimism: Be Positive

Positive thoughts induce positive results. Influencers such as Tony Robbins and Zig Ziglar, preach this invaluable lesson early on in life to effectively lead and influence. In anything we do in life, we must be positive and believe in ourselves. Therefore, the more optimistic we are, the more opportunities will open for us and those around us. Because the majority of people prefer to surround themselves with their fellow peers who possess more positivity, knowledge, and sense of happiness.

Congratulations and welcome to leadership's first trait! Optimism is the first quality people will require in distinguishing and recognizing you as a leader. Believe in yourself and the power of hopefulness. Avoid pessimism and turn negative conversations into positive ones. Galvanize people into changing their state of mind into a bright and positive place. When you accomplish what is equitable by you and what is fair to others in a positive way; people will rely on you and expect noble deeds. They will follow you, and eventually, become loyal customers.

Assuredly, I am not the first one to point this out. However, I am hoping this will be a turning point for you. Utilize optimism to enhance your life and your career positively.

"Once you replace negative thoughts with positive ones, you'll start having positive results."

—Willie Nelson

Happiness: Be Cheerful

Love yourself and do what makes you happy. Let's not confuse this statement with greed or selfishness. In the event of an in-flight landing emergency, you need to put on your mask first before helping others. Similarly, you can't make others happy if you're not! Therefore, living your life, knowing that YOU are a top priority is what will aid you to help others. Upon reaching a point where you feel content and satisfied with your lifestyle, you will feel your purpose. When you serve and support others, you will experience even more joy and satisfaction.

Occasionally, we complete tasks we're not happy with due to the fear that others are going to judge us, or feel inferior. Overall, this can result in our desire for others to be happy, but it doesn't make us happy. For example, it is effortless for us to respond with a, "Yes" when we honestly wanted to say, "No." We are taught not to turn people down! Leaders often recommend answering "yes" as often as possible, because there is an extraordinary power in doing so to achieve success. However, sometimes, we find ourselves in situations where saying "yes" doesn't feel right. How often have we said "yes" when down deep in our hearts we knew we should have just said "no?"

Unfortunately, like the majority, I have fallen victim to the "yes" or "agree" tactic. This approach may be appropriate with your family and loved ones. However, in business, leaders understand the indispensable necessity of saying "no," or they offer alternative solutions. That way, they only agree to viable requests they deem worth their time and happiness.

Further, there is a difference between making a sacrifice for a good cause and being exploited. Leaders can immediately identify these situations, and so should you. Great leaders may bend the rules, but they will not break them. They will bend them only if they believe in a prominent benefit in the long run. They will not break them just for the sake of others, and most will consider each request thoroughly before saying "yes." They strive to lead, do what's best for them, their business, and what makes them happy. Leaders stand up for themselves, their purpose, and their families. They understand what they want, have control over their business and their lives.

Run your real estate business, but don't let your business run you! Bend the rules, but don't break them! When things don't feel right, don't be afraid to say "no." Offer alternative solutions that work in the best interest of both parties. If a seller tries to negotiate your commission to an unsatisfactory rate, don't be afraid to say "no." You unquestionably can't serve everyone, and you cannot be everything to everyone, but you can always do what makes you happy.

In a nutshell, great leaders are influential individuals who understand their goals and commit to them. They aren't afraid to say "no" to situations that don't align with their purpose. Similarly, you need to realize who you are, where you stand, and what you want. Stand up for yourself. Take control of your decisions, your business, and lead your life. Do what cheers you; strive for whatever is best for you, your business, and what your purpose is in life. Be happy, and don't be afraid to say "no." Don't let others control who you are, or influence your decisions and behaviors. Once you are capable of doing what you want and what makes you happy, you are ready to become a leader. When you are in control, you will dominate whatever you put your mind to —because you are in charge and the influential leader of your own life.

Let me tell a little story about Ross. Ross was frustrated because his buyer signed with another agent, even after driving the buyer around on the weekends looking at Open House events. Ross didn't have a signed exclusive agreement! He was hesitant to ask the buyer to sign the contract because he was afraid of losing the client. Well, a buyer without an executed agreement isn't a serious buyer. Why commit to a person that neither wants to commit to you nor values your time?

The next time potential buyers ask you to show them homes or drive them around the neighborhood; ask them to sign an Exclusive Buyer Agreement with you. Here's an example script that you can use the next time someone asks you to drive them around:

- *Bob the Buyer: Hi Ross, can we drive around the neighborhood and look at Open Houses this Saturday?*
- *Ross the Realtor: Sure, Bob. I'd be delighted to show you around the neighborhood. To set up appointments with various listing agents, I have to represent you officially. Are you free later this afternoon or tomorrow at 10 AM to meet at my office so we can discuss the terms of the buyer's agreement?*

If they resist, and if you are a new agent, or if you haven't earned your buyers confidence, bend the rules, but don't break them. Take them on tour once to see if you're a good fit for one another. If so, have them sign the contract at the end of the trip.

Gratitude: Be Grateful

There are amazing people in our lives and various incredible commodities around us. We often come across them and take them for granted. We are not to blame our busy schedules, business commitments, and life obligations. Sometimes we don't realize how fast time flies. Unfortunately, people become aware of these distinct values only when they are gone or lost.

We must be grateful for what we have and what's around us at all times. We must see the good in others and bring out the greatness in them. In the end, it is not only about what we want to accomplish, but it is also more about the people with whom we want to share our achievements. Having a great life doesn't result in it being about us, but preferably spent with those whom we love and respect.

Great leaders contact many people daily to remind them of how grateful they are for having them in their lives. They consistently strive to make others feel immeasurable. Being appreciative builds strong relationships with others and results in a network of loyal and amazing people.

Be grateful to others and make them feel special. Share their stories and let the world know why they're invaluable. Thank them for being around and ask them into your life. By doing so, you invite them to share your success and happiness, and retroactively, they will be the ones who drive both.

"Talent is God-given. Be humble. Fame is man-given. Be grateful. Conceit is self-given. Be careful."
—John Wooden

Take the Next Step
Execute Tactic 1.1: Be Grateful

Your Purpose in Life

As children, our parents praised us when we performed tasks well, and scolded us when we executed duties inadequately. Perhaps there were times where we were focused on the negatives and the mistakes profoundly, and how to avoid them, that we are left oblivious in seeing the greatness in ourselves and others.

As we mature, we start learning how to improve ourselves. We diligently work for a good education so we can be our best in the workforce and establish better lives. We spend most of our years learning how we can become better. We consistently think of and examine ways to improve on what we can do and what we can achieve —in business and our personal lives. We do all of that to continually strive and find out what we are truly meant to be. It takes experience and guidance to find our purpose in life. As we grow, two types of people evolve. Most will transform into the person that wants to live and enjoy a simple life being content in one's current living situation. Some individuals wish to influence, dominate, and leave their mark in the world. Leaders are the latter.

Disclosure

By classifying the types of people below, I am in no way judging anyone. I am merely trying to get my point of view across. I want you to be able to identify the people that should surround you. It is vital that you distinguish the right people that will have a positive influence on you; whether around you or in your business. Doing so will positively affect your life and your happiness.

The Self-Centered, Pessimistic

As the proverb stands for these types of people, the glass is always half empty. They focus so much on their future that they become self-centered. They think the world revolves around them, that everyone still wants to know what's happening in their life. When things don't go their way, they complain, as if their entire world had just collapsed. Unfortunate as it is, a majority of those people become pessimists because they always have certain expectations of life that may never be enough. Have you come across these types of people? Did you sense how their adverse reaction somewhat affects your energy and puts you down? Though they may not mean it, and it may not be their fault, but somehow, their purpose in life becomes mostly self-centered, and it doesn't help your cause. These are the people that, no matter what you do, no matter how well you do it, somehow, someway, you'll end up with a negative review on "Yelp." Leaders avoid them at all costs, and so should you. If it's inevitable to avoid them in your personal life, then attempt to avoid doing business with them.

> *"A pessimist sees the difficulty in every opportunity; an optimist sees the opportunity in every difficulty."*
> —Winston Churchill

The Humble & Optimistic

On the other hand, you meet others that are down-to-earth, modest, positive, and hard workers. They support themselves and their families. Their purpose in life is to do well and contribute to the world in which they live. Money is likely highly relevant to them, because without it — they utterly can't survive. They understand the positive ways that money can contribute to enhancing different aspects of our lives. Those are the people that you want to do business with —as partners, employees, and customers. Those are the kind and most loyal customers. Those are the people that go out of their way to give you a five-star review. Leaders seek and favor them. They try to recruit them into their lives, and so should you.

"The quality of our lives is as great as the quality of who we are and those around us."
—James Tyler

The Successful & Aggressive

Some people are business-minded executives. They become entrepreneurs; they build profitable companies with the mission to generate millions in revenue so they can live luxurious lives. They are quick on their feet, aggressive and assertive. Those tend to be leaders and the most successful ones. Part of their purpose in life is to create wealth —a whole lot of it. Those are the people with whom you want to be associated. Those are your diamonds in the rough that might require extra care and attention, but they are your best clients. Find them and hold on to them —as clients and partners!

The Legacy Leaders

While food, water, oxygen, and money are essential parts of life, they are not the reason why we live. A great leader must change lives and serve a noble purpose that is bigger than revenue. Leaders strive to make a meaningful impact on someone's life and a noticeable difference in the world. Thus, they've figured out WHY they exist. By helping others, they are likely to make money —perhaps tons of it. If you start with the sole desire to make money, you won't make an impactful change. To make money, you must commence with the desire to make a difference. You need to think and act like an entrepreneur, a business owner, and a legacy leader. Most importantly, you must know why you are here.

The greatest leaders understand that they can't please everyone, but they find ways to make effective contributions and a distinctive impact on people. While there is nothing wrong with selfish motives, as a legacy leader, you will need to be influential. You will need to go beyond yourself and captivate the imagination of other people. You will need to clarify your values, know your purpose in life, and have a vision that paints a fantastic future to help others.

If you want to be a legacy leader and dominate the real estate marketplace, I must ask: *"How will you positively change other people's lives?"*

It might be many people or just a few. Use your imagination and figure out how to positively impact people's lives. How does your presence here on earth make people's lives better and different? What is your purpose? Will your answer be a cause that will make people believe in you, your legacy, and follow you? This question is not something I can decide for you. It is the one and the most significant decision you need to choose for yourself. Build your life; not just your wealth!

> *"The purpose of great leaders is making our world a better place, each in a particular way; find yours!"*
> —James Tyler

Your Actions

I love Malcolm Gladwell's quote in his book Outliers, *"No one who can rise before dawn three hundred sixty days a year fails to make his family rich."* It is so authentic and authoritative. Are you a go-getter who takes daily actions that progress towards a meaningful change? How are you positively changing your life and those around you?

Take Initiative

Take the initiative, take charge, act, or start —is the first tangible step you take in your leadership and success journey —whether in your business or your life. Taking the initiative creates momentum that overcomes all fears. It is the passion that drives your ambitions and your motives. It is what matters most in achieving results.

Great optimistic leaders are full of energy. You can always feel their enthusiasm. They strive to overcome the top challenges with endurance and determination. They are tenacious and resilient. They build enterprises from the ground up with persistence and perseverance. You can do it too.

Fear is the number one factor that prevents most real estate professionals from thriving and accomplishing outstanding results. Whatever your concerns are, overcome them today by taking action. If you don't fail, you won't learn.

This book started with a word, a sentence, and then a page, with many challenges —in thinking, writing, and overwhelming thoughts. I am like you; I was also afraid; can I do this? What if I mess up? Will people judge me? Despite it all, I took charge and made it happen! Remember this; your next listing could be just a phone call away! The only thing that is standing between you and that deal is you.

Pick yourself up, stand tall, and get out there. Have the grit to dominate. Start and end your day with a ritual. Whatever you wanted it to be —for example, *"I fear none. I can do this, and I will."* It should be the first thing you say when you wake up in the morning, and the last words you utter when you go to sleep: *"I fear none. I can do this, and I will."*

Whatever it is that you want to accomplish in life, whatever challenges come your way, you must believe that you can. Once you believe in yourself and your abilities, you will take the initiative to lead, dominate, and achieve your desires. Create a ritual now! Use the following three E's to take action and get to wherever you want to be in life.

The 3 E's for Taking Initiative

To sum up the characteristics of initiative in great leaders, I would use *Energy, Enthusiasm,* and *Endurance.* You need a vigorous effort to build a successful real estate enterprise. Effort springs from *energy, enthusiasm,* and *endurance!* Be bold, take the initiative now, and put yourself in motion. Think of your wants and needs. Start crafting them into tangible goals, and execute them to fulfill your desires. You are always one step away from making that drastic change in your business and your life. You never know which level is the trigger to that big step. Become a person of action and lead by example. Leaders don't always know what to do; they do. They learn along the way, and they do better the next time. The vital key here is to take the initiative and act immediately!

> *"You miss 100% of the shots you don't take."*
> —Wayne Gretzky

IMPORTANT:

Access the "Dominate Real Estate Methods & Tactics" at https://Dominate.RealEstate

Take the Next Step

Execute Tactic 1.2: Take Initiative 1.0
Execute Tactic 1.2: Take Initiative 2.0

PART II

MANAGEMENT

CHAPTER TWO

ARTICULATE DESIRES INTO TANGIBLE GOALS

"If you don't know where you are going, any road will get you there."
—Lewis Carroll

This chapter is designed to help you plan, execute, measure, and track your success. You will learn how to put your desires and wants into attainable goals that become practical plans to implement and achieve them.

Plans require objectives that are tasked into daily activities to either be handled by you or delegated to someone else. With growth-driven goals, you can build a powerful growth system. However, these objectives must leverage time, people, and technology.

Furthermore, you must practically implement empirical indicators to measure your real estate success. These indicators will drive the system forward based on its performance. With this information, you will identify areas of opportunities with goals and objectives driven towards growth.

Set Your Goals

There are no secrets to goal setting. I bet you've heard and read this many times before: your goals must be SMART: Specific, Measurable, Achievable, Relevant, and Timely.

Besides having specific goals with a clear finish line, I challenge the criteria of this acronym. Most businesses find that the SMART model isn't sufficient to drive them into taking action. Your goals must also be so inspiring with a purpose that will continuously drive your momentum forward, even after achieving them. How many SMART goals do you know of that you did not accomplish? They were missing the stimulating factor that keeps the thrust going way past the finish line.

Therefore, I have added three additional factors to the SMART concept—*Tensile, Tactical* and *Thrilling*, making it SMARTTTT or **SMART²**.

- **Tensile:** Your goal must be scalable and flexible to stretch. Numerous studies have found that committing to ambitious; seemingly-out-of-reach goals can spur incredible results in innovation and productivity. An identical theory is called "big, hairy audacious goals," or BHAGS, proposed by Jim Collins and Jerry I. Porras in their book entitled Built to Last: Successful Habits of Visionary Companies. If you find yourself easily and quickly accomplishing your goal, can you expand the target so you can live up to your highest potential?

- **Tactical:** Your goal must be diplomatic, nimble-footed, and agile in the strategy where incremental achievements create impactful meaning. It would help if you tactically designed your goals around your vision with small wins and micro-victories to achieve the ultimate goal. A goal can have many tactical elements. For example, can you celebrate the 25%, 50%, and 75% marks in your journey towards achieving your goal?

- **Thrilling:** Your goal must be compelling with an added emotional factor. It must evoke interest, stimulate exhilarating energy, and provoke excitement. It must directly connect with your vision of what kind of business you want, or what type of person you see yourself becoming. For instance, are you excited every morning to rush out of bed and work on fulfilling your goal?

Let's compare a SMART goal to a SMART² goal:

- **SMART Goal:** *I want to close 120 clients in city x in the next 12 months.*

- **SMART² Goal:** *I want to acquire 12,000 contacts in the next 60 days, from city x, so I can start closing 10 listings per month, consistently, starting 90 days from now.*

Both examples align with the five characteristics of the SMART concept. The goal is to close ten listings per month. However, in the latter option, obtaining the 12,000 contacts is a concise tactical factor that is needed to fulfill the strategy. It is a micro-accomplishment that gets you halfway through. Think of it as a micro-goal. Even if you didn't close ten deals per month; getting those contacts would be quite an achievement in itself. That said, every goal needs a tactical element to act as a stepping stone in your journey to reaching the final destination.

When I point out the 12,000 contacts, it makes the ten listings look like a breeze to achieve. It will; because you already did the math. Can you feel the thrill and the momentum behind the consistency of closing ten listings every month due to having 12,000 contacts in your database?

You may have come across coaches where they advise you to go all in and set up 10X goals. 10X goals refer to setting goals 10 times greater than what you think you can attain so that you can push yourself to reach somewhat satisfactory results. In Grant Cardone's book, *"The 10X Rule,"* he says that 99% of the people in the world think that they are incapable of achieving great things, so they aim for the mediocre. While I agree with Grant about not underestimating your abilities, I cannot entirely agree with the fact that you have to set up unusually lofty goals. It wouldn't be wise to have unreasonable targets only to infuse adrenaline into your actions, then fail and get discouraged. Instead, it would help if you were enthusiastic about achieving your goals, regardless of their size. For instance, would you set a goal to close 30 sales a month if you were merely starting your business with a limited budget? Be careful of the hype and play within your means. Although, do not set goals where you can swiftly achieve and go without any noticeable results.

What I have found best in setting and accomplishing goals is always to have three goals:

1. Short-Term Goals (one month to one year)
2. Mid-Term Goals (one year to five years)
3. Long-Term Goals (five to ten years plus)

Short-Term Goals

A short-term goal is to achieve immediate results that are urgent and important. For example, you can set up a financial goal to produce active income, which you can call the **Survival Income Goal**. In this case, it would be more reasonable to set up a goal where you can thrive on making what is needed to pay your bills. It would be devastating to plan for creating an excessive amount of money, where you risk not achieving anything, and your bills go unpaid.

Mid-Term Goals

A mid-term goal would be to produce significant, but not instant results. For example, you can set up another type of financial goal to generate passive income and steady cash flow. You may want to title this goal your *Steady Income Goal*. When planning the mid-term objective, aim to save money and live comfortably without having to worry about paying your monthly bills.

With a recurring *Steady Income Goal* in mind, you will focus on the urgent and vital short-term goal, but simultaneously striving towards the mid-term goal to fulfill a stable and steady income.

Long-Term Goals

Finally, your long-term goal is the 10X goal. This goal gets you super excited, and the goal that you think will make you the happiest person on earth! This one is going to pay off in ten plus years, and the one that is supposed to be extremely difficult to accomplish. You could also set this to be another financial goal to produce retirement income, and call it the *Secure Income Goal*, the seclusion income goal, or the $100,000 per month income goal. Whatever you want to call it, know this is the goal that motivates leaders, fires up their emotions, and drives their momentum daily. It is not the goal that you want to accomplish tomorrow, but it's the legacy goal that motivates you daily, and fuels your energy with tremendous enthusiasm.

Since people commonly associate goals with being financial, in my example, I provided three financial goals. However, I highly recommend following the same Three-Goals-Concept in setting other types of goals. You can set up different goals pertaining to your personal life, physical health, and productivity. In doing so, follow the level of urgency and importance to prioritize execution.

1. Short-Term Goal: Define what is urgent and vital.
2. Mid-Term Goal: Define what is not urgent but necessary.
3. Long-Term Goal: Define what is not critical and not presently essential but highly motivating.

Set up your three SMART² goals:

1. **Short-Term Goal**: *I want to acquire 12,000 contacts in the next 60 days so I can start closing 10 listings per month, beginning 90 days from now.*
2. **Mid-Term Goal**: *I want to hire a team of 50 agents, 10 per year, for*

the next five years.

3. **Long-Term Goal:** *I want to become a real estate investor who owns ten apartment buildings in the next ten years.*

In setting your goals, use the "Put First Things First" methodology explained in *The 7 Habits of Highly Effective People* written by Stephen Covey. Covey talks about the difference between leadership and management. It is the same method I use to set up my priorities. It is a matrix of *"importance"* vs. *"urgency"* that Stephen Covey and Dwight Eisenhower used in deciding where to invest their efforts. Priorities are split into four quadrants and assigned based on the order of *"urgency"* and *"importance."*

1. **Quadrant I:** Do the "urgent" and "important."
2. **Quadrant II:** Plan the "not urgent" but "important."
3. **Quadrant III:** Delegate the "urgent" but "not important."
4. **Quadrant IV:** Eliminate the "not urgent" and "not important."

Quadrants one, two, and three determine High, Medium, and Low priority levels, respectively. For a better grasp on time management and priority settings, I highly recommend reading *"The 7 Habits of Highly Effective People"* by Stephen Covey.

Go back to the drawing board if you are in the initial stages of setting your goals, and you are not clear about the essential steps in prioritizing your objectives. If that's the case, here are the three critical steps that can help you draft a vision for setting your goals and establishing their priorities:

1. **Create a vision board:** This could be a whiteboard at your office, a Trello board, or sticky notes on your bedroom mirror. It doesn't matter where you jot down your vision as long as you can easily access it and revisit it daily. Your view cannot be in your imaginary brain. It has to be written in an accessible place, hence, the vision board. This way, you'll stay focused on the big picture and pivot if and when necessary.

2. **Feel your vision:** Align the reality of your vision with your emotions. Fast forward and see your vision accomplished. Your image is now a reality, and you are living that life. How do you feel? Your emotions will tell you whether you have the right vision or still need some tweaking. The combination of visualization and feelings will help you see a precise image and eventually set concise goals. Otherwise, you will be working hard at vague objectives and uncertainty.

3. **Foresee anticipated roadblocks**: Discover the hidden barriers that might prevent you from achieving your vision and accomplishing your goals. Prepare yourself to face those hurdles and position yourself to overcome them, starting with the most difficult ones first. Hustle and pick yourself before having to deal with unfavorable situations at the most unexpected times.

Goal Types

When you set your goals, it is necessary to focus beyond your real estate goals. When people think of future success, they often immediately associate their goals with their business and automatically think of ways to improve their career. While lifestyle and wealth are essential; in reality, there are other primary goals for different aspects of life to set.

Defining who you are and what makes you happy goes beyond just wealth and your business goals. Understand the emotions that will fulfill the inner you so you can have a better vision for setting your goals. What kind of feelings will a better version of "you" experience? What are your passions? What would completely satisfy you as a person? Think of your physical health, personal interests, your family, friends, and perhaps your spiritual and emotional desires.

Satisfying your passion and emotions is very healthy and rewarding. Being mentally sharp and physically fit improves productivity, confidence, and self-esteem. Some people once finished with formal education, never reread a book or think of pursuing continuing education. Set goals to learn new trades or hobbies continually. Goals don't always have to be about business. They can be personal and about pleasure as well.

Life is continually evolving; and so should everyone. Don't focus just on business and wealth; otherwise, you might suffer to find happiness by the time you accomplish those business goals. Keep your life top of mind and imagine a bigger vision for your life; seek to be well-rounded and more effective in each area of your life. Set goals to create your happiness regardless of business requirements. Control your life because inevitably, the quality of your career will enhance by improving what matters most to you.

The 5 P's of Goal Type

Here are the Five P's of Goal Types that will transform your life and those around you:

1. **Psychological**: Mental attitude, mindset, emotions, prayer, and meditation.
2. **Philosophical**: Education, knowledge, self-improvement, and personal development.
3. **Physical**: Physical activity, sports, medical self-care, nutrition, and rest.
4. **Personal**: Lifestyle, family, friends, relationships, travel, and hobbies.
5. **Professional**: Business, career, and wealth.

Being aware of these five elements in your life and contributing to each one daily, weekly, or monthly will have a significant impact on your happiness. Regardless of the goal type, the merit is to have the right mindset towards setting your goals in life. How you think about life in setting your goals is critical to your overall success.

Your mindset affects every area of your life and can determine your level of motivation toward anything you set to achieve. Thus, in real estate, you have to have reasonably high goals with a mindset to grow. Become 100% emotionally invested in making a positive change in both your life and others. Otherwise, it will be very challenging to succeed.

In the end, you can make all the money in the world; however, when you leave earth, you take nothing with you but good deeds. Rings true to what Billy Graham once said: *"The Brinks Truck Doesn't Follow the Hearse!"* Hence, envision what makes you happy. When you get there, come back to reality, and set your goals wisely to get there.

"Build Your Health; Not Only Your Wealth!"
—James Tyler

Take the Next Step
Execute Tactic 2.0: Set Your Goals

Execute Your Goals

While creative vision, clarity, focus, commitment, dedication, enthusiasm, and many other factors are required to execute and achieve your goals, the three most important determinants are leverage, execution, and accountability.

Leverage and Time Management

Now that you have set up your goals, you need to plan out the steps to execute them. However, how do you know what you should do first? First of all, you must understand that you should not have to do everything yourself. Second, you need to know the difference between high leverage activities (HLAs) and low leverage activities (LLAs).

A high leverage activity, sometimes referred to as a high value activity (HVA), is a task that only you can do. The outcome of executing such action has a direct impact on your business, wealth, or life. Knowing how to perform something that doesn't directly and positively impact your life doesn't qualify as a high leverage activity. For instance, writing a blog article doesn't qualify as a high leverage activity. You can hire someone else to do so. However, recording a personalized video to a potential client does.

On the contrary, a low leverage activity, also known as a low value activity (LVA), is defined as an activity with little importance. It is a task that has a small impact on your business and your life.

High leverage activities get you the results you need faster, with less expenditure of time, money, energy, and talent. Conversely, low leverage activities consume resources and drive results more slowly.

Top real estate agents focus on HLAs that most positively influence their business. They leverage time, people, and technology to improve productivity, accelerate achievements, and increase profits.

To become a top agent, identify significant HLAs that give you the best advantages in your marketplace. Meet with qualified clients to get signed contracts and delegate LLAs to others. Strategically manage your time and leverage other people to achieve more. Determine what your time is worth, and then take advantage of *"The 3 T's Ladder of Leverage: Time, Tribes, and Technology."*

Leverage Time

Understanding how the most successful real estate professionals manage their time is crucial to your success. Before looking into time management hacks, let us make sure you calculate what your time is worth first.

Let us assume you want to get paid $200K per year, and you want to work 40 hours per week. Assuming you want to take a 2-week vacation, you'd be working 50 weeks per year. That is 2000 hours per year. Your hourly rate would be $100 ($200,000 divided by 2000 hours). Similarly, to make $500K per year, you must be doing tasks that are at least worth $250 an hour.

To grow and dominate your real estate marketplace at $500K target annual income, you must have a dynamic business model. You must build a robust system that will attract, nurture, and convert people to loyal customers. The most critical factor is over 90% of your system's structure must be built by others who will cost you less than $100, and indeed less than $250 per hour.

Most real estate agents face the challenge of making $250 an hour because they spend most of their time trying to accomplish tasks that are less than $50 an hour. Do not lessen the chance of scaling your business by investing your precious time in LLAs instead of your money. The other challenge other agents have merely is not knowing the steps necessary to get themselves in a position to focus on the $250, $500, or even the $1000 per hour tasks.

Gain more leverage and better results from your time by prioritizing all tasks at all levels. Do the things that get you paid the big bucks. For example, invest in learning new tactics that have a direct influence on your business and your life. Hire two or three individuals at once and mentor them in groups since one hour of mentoring will likely pay for itself a hundred times over. Build a dynamic sales process and automate repetitive work. The results of these crucial high leverage activities are worth a multiple of the energy invested into them. They have a direct impact on maximizing your productivity and the outcome of accomplishing your ultimate goals.

Leverage Tribes

By "*Tribes*," I mean people in your network. Tribes could be family, friends, clients, employees, partners, followers, fans, and external sources such as freelancers and virtual assistants.

Large enterprises create Standard Operating Procedures (SOPs) to achieve efficiency and increase productivity. Why not create similar instructions within your real estate business? To leverage people within your work environment and clients outside your office, create a "Standard Rules of Engagement (ROE)" policy. For instance, if you create a rule that you will do listings in the evenings and weekends only. Then, you will create another command that you will meet with new clients only during weekdays and between 9 AM to 6 PM. Also, if you consider hiring a real estate executive to meet up with potential new clients, qualify them, and evaluate their situation. You would create a complementing rule that you will only meet with those who are eligible for the next step, whether it is to tour the house or conduct a listing presentation. Further, you may want to hire another sales representative to fulfill the rule of scheduling and managing listing appointments.

Leveraging a real estate executive to qualify prospects, and a sales representative to manage your calendar frees up time for you to focus on high leverage activities.

Again, leverage people by understanding your salary, and employing those with a less hourly rate than yours to do the low leverage activities. For instance:

- If you want to make $50K annually, hire people to complete your low average activities for less than $25 hourly rate.
- If you want to make $100K, hire people with less than $50 per hour
- $200K per year, hire people with less than $100 per hour
- $500K, less than $250 per hour
- $1 Million, less than $500 per hour.

Based on your available budget and the annual salary you realistically set for yourself, it would be best to leverage time and people. Delegate the easy tasks, and those are less than your pay rate. *Insource* or *outsource* all administrative duties and any other functions that don't affect your brand reputation. One exception is not to outsource *sales* or *customer service*. Also, anytime you find yourself doing something repeatedly; it will either need to be *automated* or *delegated*.

Leverage people's time and help so that you're not trying to finish everything yourself. You want to remove yourself from any mundane or non-critical tasks eventually. Discipline yourself in creating a paradigm shift. Build a "To Delegate" list before you create your "To Do" list. You want to focus on substantial, essential opportunities that are highly critical to your life and your success. Invest your energy and strengths in the most difficult, strategically, and rewarding tasks—delegate the rest.

Optimize your production by following *Parkinson's Law* and the *Pareto Principle*. Less isn't laziness; schedule less time to do more. *"If you wait until the last minute, it only takes a minute to do," "work expands to fill the time available for its completion,"* and *"roughly 80% of the effects come from 20% of the causes."* Adhere to those due dates you put on your execution plan because 80% of your results are 20% of your efforts. Avoid procrastination and maximize your productivity by optimally leveraging your time and the time of others.

Leverage Technology

Get more done in less time. Being organized and efficient is one of the most critical factors of success. Leverage available tools and technologies in managing your daily activities and collaborating with others working on your business.

Whether you are building a website, running marketing campaigns, or following up with clients; use management platforms such as Trello or Asana to organize your projects and daily activities.

Another part of collaboration is effective communication. Manage communication within the entire office and with your team internally. Use online chat applications to discuss daily activities, your clients, and your office needs:

- If you have Google Suite, use Google Chat: *https://chat.google.com*
- Otherwise, use Slack: *https://slack.com*

These platforms are available on the web and in mobile applications. I recommend you take advantage of online chat applications to save time and improve productivity performing high leverage activities only.

Take the Next Step

Execute Tactic 2.1: Identify Your High versus Low Leverage Activities

The Execution Plan

Execution plans, also known as action plans, are the most underrated and under-used strategies in the real estate industry. We are going to change this. Well, at least for you!

Now that you have your three goals, the only goal that you need to focus on is the short-term goal. That is the urgent and most pressing one. To accomplish its objectives, you need to envision the final destination, plan out the steps that must be taken to get there, then execute them.

Great leaders are doers, but before they execute, they plan. That is what we call, an "execution plan." The plan must be interactive, where it keeps the process exciting. It must have milestones. These are triggers implemented within the method to check results against expectations periodically. Effective executives usually build a minimum of two milestones into their execution plans. They determine and measure fractional achievements. Besides, they test whether they are meeting desired outcomes, or if future revisions are needed. The first breakthrough comes halfway through the plan's period. The second occurs at the end, and before the next execution plan is drawn up. Milestones are required to re-examine the strategy as events unfold. Otherwise, it will be challenging to determine which activities, objectives, or tasks were most effective, and which were just noise.

The execution plan is more of a statement of intentions rather than a substantial commitment. It might need to be revised now and then because every success could create a need for a shift, as does every failure. The same is true for abrupt changes in the industry, the market, and the business environment. A written plan should anticipate the need for flexibility and improvement.

Reaching milestones within the plan ensures that we are on the right path and tracks how far we are from the finish line. These triggers keep us agile and energetic in our efforts. Over time, they help us overcome more significant fears and enforce habitual disciplines.

If not reached, then a need to pivot your efforts and shift your focus on what will produce better results is necessary. The key is to stay persistent, keep on going, and never give up until attainment of your goals.

> **Take the Next Step**
>
> *Execute Tactic 2.2: Get a New Real Estate Deal within 30 Days*
> *Execute Tactic 2.3: Execute Your Plan*

Accountability

To hold yourself accountable, you must form a habit that becomes your accountability program. Focus on what you want. Make it your one very most important goal. Make that goal your morning and weekly ritual. Put that goal on your to do and to delegate lists. Also, on your calendar and your execution plan. **If you do not schedule it, it doesn't exist**. Execute on it and consistently follow up.

Turn your goals into weekly milestones, then into daily actions, then into a routine. Do the fundamentals over and over, month after month, year after year, again and again. Take responsibility for your actions, commit to the dedication and hard work, daily. Those morning rituals will become your actions, and your efforts will become your routine. Also, your routine will become a habit, and your habits will become your accountability program.

Be determined and relentless. Know that you might fail, many times! Have resolute perseverance to never give up; neither daily nor continuously over the years. Stay on the course, and keep in mind that the one thing that we don't notice and see daily is "change." Change happens over time. Without accountability, it will be very challenging to see a noticeable difference, especially a positive one.

As a business person, ask yourself daily: *"What can I contribute today that will significantly and positively affect my performance and achievements towards my goal?"* Do not make a promise to yourself or anyone else that you are not going to deliver. Do what you say you are going to do.

Let's look at an example to relate accountability to your real estate business.

You are driving, and you get a call:

- *Sally the Seller: "Hi Rose, I think I want to sell my house, but I don't know what it is worth."*
- *Rose the Realtor: "With pleasure, Sally, I will look that up and get back to you with some comparables."*

Rose gets on other phone calls, and once she gets to the office, she responds to emails. Her mind wanders in a hundred different directions, and she forgets to call Sally the Seller back. One or two weeks go by, and it hits her—she got a phone call from Sally. Rose quickly sends her some comparables and calls her, but then it's too late. Sally is now working with a different agent. She thought Rose wasn't interested or was a flake. Regardless, Rose can't recover the situation because Sally lost her trust in her; that if there was one. Rose didn't do what she said she was going to do — get back to Sally with some comps.

Without an accountability program, small mistakes like these cost agents thousands of dollars, relationships, and future business—top producers are no exception. What do you do to avoid something like this from happening to you? Simple! Develop a system to hold yourself accountable, or ask someone else on your team to hold you responsible. Immediately ask potential customers to send you a text message or an email with a piece of information, i.e., their property address. This way, they will not only remind you but also show how serious they are about working with you. Immediately put a reminder on your calendar. Also, add the task to your to do or to delegate list. It is imperative that you implement a method that participates in reminding you of unfinished activities calling for your attention. Don't miss reconnecting with your clients and following up promptly.

Here are some ideas that you can hold yourself accountable to achieving:

1. Write one or two guides that educate your buyers and sellers.
2. Write an eBook on "The Top 10 Real Estate Tips for Sellers."
3. Write an eBook about "How the Home Buying Process Works."

4. Write another eBook about "How the Home Selling Process Works."
5. Create training material that will motivate your customers to do business with you.
6. Create an online blog or online real estate course for buyers and sellers.
7. Curate the top 100 trending and most impressive real estate blog articles into one white paper.
8. Schedule 1,000 social media posts for this year.
9. Plan on making 1,000 phone calls this year.
10. Create ten live Facebook, Instagram, or YouTube videos in the next 30 days.

Tip

If you want to take accountability to the next level, I highly recommend reading The 4 Disciplines of Execution, by Chris McChesney and Sean Covey. Especially if you are a real estate broker managing multiple teams of agents. These authors did a great job building a reliable system that teaches you how to be accountable and execute on your goals effectively.

Take the Next Step

Execute Tactic 2.4: Follow up to Get a New Real Estate Deal within 30 Days

Measure Your Performance

Every goal in your business must be measurable. By regularly measuring your results against your objectives, you can identify what tactics work well, and adjust those that don't yield the desired results.

Executing an existing plan and periodically measuring against its goals is almost non-existent in real estate. The next time you meet your fellow real estate agents, ask them the following: *"When was the last time you reviewed your execution plan? Also, how far are you from reaching your goal?"* Do you feel the silence? Lucky for you, this won't be an issue moving forward. Let's dig in!

Key Performance Indicators

Key Performance Indicators (KPIs) are critical indicators that measure the effectiveness of advancement toward achieving an intended goal.

They are used to create an analytical basis to focus your attention on what matters most. They provide objective evidence of progress to help inform better decision-making that will improve performance and achieve your goals.

Many types of key performance indicators vary based on departments, industries, and roles. Here are the ones that I deem necessary for you to be familiar with as they will come up in future chapters.

- The top KPIs used in marketing metrics (i.e. Marketing Qualified Leads or MQLs, Sales Qualified Leads or SQLs, Life Time Value or LTV, Customer Acquisition Cost or CAC, and more).
- The top KPIs used in web metrics (Traffic Source, Click-Through Ratio or CTR, Lead Conversion Rate, and more).
- The most commonly used KPIs in sales metrics (Calls per Sales Rep, Monthly Sales Growth, Opportunity-to-Close Ratio or Sales Closing Ratio, and more).
- The best social media marketing metrics (Audience Base, Audience Reach, Audience Engagement, Traffic Conversions, Followers Growth, Social Shares, and more).

Suppose you want to generate leads using Facebook. You decide to write an eBook, place an ad on Facebook, and generate buyer leads. The potential audience size is your "Base." That is the first indicator you will need to evaluate. You run your ad, and now you're tracking the reach of the post through Facebook Insights. "Reach" is the second indicator you will need to trace. Next, you analyze the engagement such as likes, comments, and shares. "Engagement" is the third indicator. You can also track social visitors, Click-Through Rate (CTR), and other metrics. The one KPI that matters most is the number of "Traffic Conversions," in this case, the number of eBook downloads or leads acquired.

There are hundreds of KPIs and metrics that can be used to track performance. To clearly articulate your performance progress, and have a better understanding of your performance levels; take the time to establish specific performance baselines, targets, benchmarks, and standards. This process is called performance management.

Performance Management

Performance management is a cycle implemented to tell you how well you are doing in reaching your objectives and fulfilling your goals. It consists of four performance measures, each to help you manage the use of KPIs to have a better understanding of your performance towards achieving your goals. The four performance management measures are

Performance Baseline, Performance Benchmark, Performance Target, and Performance Standard.

Performance Baseline

A performance baseline is the starting point measure, or "The Start Line." It could be the average level of performance where you are currently, which could be closing one listing a month. Use the performance baseline measure to compare your future performance levels and test how your performance is changing over time. For instance, you can set a performance baseline of acquiring and closing a minimum of two real estate deals a month.

Performance Benchmark

Performance benchmarks are typically measures from outside your company. They are levels of performance that other competing organizations have achieved. It would be in your best interest to reach the milestone because others have achieved that level of performance. For instance, the benchmark standard is on average, closing one real estate transaction, per agent, per month. Generally, the benchmark should be your next objective after you achieve your baseline goal.

Performance Target

The performance target is the desired level of performance that we aim to achieve in the future. Benchmarks from other organizations that represent industry best practices can become measures that you use for your performance target measures. Performance targets help track your current progress against the result, or "The Finish Line" you intend to reach. Let's say your aim becomes consistently closing two real estate transactions per month.

Performance Standard

Performance standards are industry-established values. They are realistic and approved measures of the performance threshold expected. For example, the National Association of REALTORS® (NAR) reported in 2017 that the typical agent has, on average, 12 transactions per year. This performance standard is one transaction per month. So, if your benchmark is currently zero listings per month, you need to get to one listing to match the performance standard in the industry.

If you want to dominate, you need to start with the performance standard or performance benchmark, whichever is higher. If your benchmark is higher—for example, you are already closing three listings per month—this measure becomes your performance baseline. Your performance target now needs to be higher than three listings per month. Making this goal will keep you ahead of the average real estate agent, at all times. Here's an example:

Suppose your performance baseline measure is a consistent close of three listings per month. Other real estate professionals in your farm close an average of four listings per month. Since you are one listing below the benchmark, your performance target will need to be set to seal a minimum of four listings per month. Once you meet this benchmark, which is the closing of four consistent listings per month, this will become your performance baseline. A new target that is higher than the benchmark of four sales will need to be set and met. Let's say you set a goal of six listings per month, which will then become your baseline once met. You will then be ahead of your competition. A new target will need to be set and met. You will set it to ten listings, and so on. By doing so, and regularly managing your performance, you'll not only compete, you'll dominate.

Lead Measures

Managing performance often means working to improve leading indicators that will later drive lagging benefits. Leading indicators are precursors of future success. Leading measures or leading key indicators are predictive and influenceable data inputs. Predictive is the most impactful indicator that will drive achievement, and influenceable implies the power of performance to move towards success.

Leading indicators must be worth measuring, whether once or in an ongoing process. The key is to implement lead measures and launch quickly. This way, you can test and determine critical KPIs that influence results, positively and adversely, so you can optimize and repeat.

To develop specific measures to track what is relevant in reaching your goal, you must determine the type of rules that are essential for each objective or test. There are two types of leading indicators for managing your KPIs used to measure performance: *Quantitative* and *Qualitative*.

Quantitative

Quantitative lead measures are numerical data or values that grow or decline over time. Quantitative data is the type of data which can be measured and expressed numerically. Quantitative lead measures come in a statistical format and structured level of occurrence. They answer questions such as *"How many?"* or *"How much?"*

Quantitative market research is conducted using methods like surveys, interviews, experiments, and observations, which are best at collecting structured data. Quantitative data helps in testing hypotheses, attends to precise measurements and objective data collection, and seeks a single truth as it uses reliable and valid data.

A good way to think about quantitative metrics is through the framework I shared previously on Base, Reach, Engagement, and Conversion. For example:

- **Base**: 1,000,000
- **Reach**: 100,000
- **Engagement**: 100 Likes, 20 Comments, and 1 Share
- **Conversion**: 50 Downloads

Qualitative

Qualitative measurement is a more nuanced approach to collecting insight. The classification of its data objects bases attributes and properties that are in words or text instead of numeric numbers. Qualitative data is non-statistical and unstructured. It provides patterns that answer the question, *"Why?"*

The validity of qualitative research refers to trustworthiness or credibility. Conventional methods of assessing efficacy include consistency checks gathered from recording data via a variety of techniques such as focus groups, observations, interviews, survey questions, journals, or existing documents, among other sources.

Data is sometimes complicated and has multiple realities. The goal of qualitative data analysis is to uncover emerging themes, patterns, concepts, and insights. Qualitative measures are used to understand an underlying process; that is, a sequence of events or constructs and how they relate.

An excellent way to think about qualitative metrics is through a survey. Suppose you were to send an inquiry to the fifty people who downloaded your eBook. Your request could have the following questions:

1. Did you find the process of downloading the eBook easy or difficult?
2. Why did you download the eBook?
3. Did you find the information in the eBook helpful?

4. Why was the information helpful?
5. Will you download another eBook?

As you can see, these questions are aimed at testing engagement and speak to quality versus quantity. Answers may vary per each person. You will have to manually analyze the data to measure performance and improve engagement levels.

A workaround for this method would be to convert qualitative text answers into quantitative numerical values using scale methods such as Dichotomous Scales, Semantic Differential Scales, Likert Scales, or Rating Scales. For example, *"On a scale of 1 to 10"* makes qualitative data more accessible to analyze since the scale has ten as a maximum score and one as a minimum. Ten denotes a high-quality attribute, and one being the lowest quality attribute.

If you were to set a goal: *Increase listing transactions from five per month to twenty-five per month within the next four months.* Your objective would be: *Increase production by twenty more listings in four months.* Which means your baseline is *five* listings and your performance target is *twenty-five* listings. Your *quantitative lead measures* would be something like this:

1. Increase daily calls by 20% from existing x number of calls.
2. Schedule two additional events per week in the local community.
3. Book ten other appointments per week.

Your *qualitative lead measures* would be something like this:

1. Improve call tracking quality by subscribing to a toll-free recorded line.
2. Speed up the contract processing time by signing up for an electronic signing service.
3. Improve customer service by surveying past and existing clients.

Your goals could be financial, such as increasing revenues, profits, investment performance, cash flow, or cost savings. Your goals could be qualitative, such as efficiency gains, reducing cycle times, or improving productivity. Also, your goals could be strategic, such as competitive advantages, potential opportunities, or threats to cut.

Congratulations! Now you know how to set up your goals, execute them, and measure your performance with confidence.

Let's revisit the exercise provided in the execution plan *"Tactic 2.2: Get a New Real Estate Deal in 30 Days!"*

Are you able to put this lead measure in a scoreboard to track your performance? Let's draft the qualitative and quantitative KPIs so you can measure your efforts and reach your goal.

Quantitative KPIs:

1. The Number of Calls
2. The Number of Opportunities
3. The Number of Follow-Up
4. The Number of Referrals
5. The Number of Appointments
6. The Number of Follow-Through
7. The Number of Listings

Qualitative KPIs:

1. Did you get a hold of the person? (Yes, No)
2. Was the phone number accurate? (Yes, No)
3. How would you rate the tone of voice? (Unpleasant, Neutral, Pleasant)
4. How would you measure the overall call? (Unsatisfied, Neutral, Satisfied)
5. Which communications channels did they prefer to use? (Social, Phone, Email, SMS, Face-to-Face)

Track Your Success

Players keep their eyes on the score. Raising the bar, hustling and winning the game is measured by the final score. Every sport in the world has a scoreboard that keeps track of one's performance and team scores. Otherwise, how will players stay motivated and ultimately celebrate their achievements? Similarly, how do you keep your eyes on your performance to track the success of your real estate business?

Scoreboard

To keep the momentum and enthusiasm going at their highest level, you need to celebrate your accomplishments weekly. To do so, you'll need to know your numbers.

Establish reporting intervals that make sense for your strategy. Analyze KPIs and achievement variations so you can gain a clear picture of your success. Look at the trend over weeks to make a practical assessment of your performance. Create a sales scoreboard to review your selected important metrics continually, and track them in real-time, if not daily, at least weekly.

KPI management on a scoreboard will help you determine your strengths and weaknesses. It will give your entire organization insight into your current performance. Most importantly, it will keep you motivated, drive your business forward, and dominate through clarity and effectiveness.

1. How do you or your real estate office manage your sales pipeline?
2. Can you quickly see each stage for every real estate deal in your business?
3. Do you have enough deals to achieve your goals?
4. Which client needs your immediate attention?
5. When do you celebrate your next closing?

Having an at-a-glance statistical board to measure achievement and progress toward a particular goal is a must. Create a cloud-based sales scoreboard using your Customer Relationship Management system (CRM), Google Sheets, or iCloud Numbers. Consider answering the above five questions while creating your scoreboard. Also, I encourage having a physical whiteboard at your office. This way, you will have access to your numbers daily at work and out in the field.

- Google Sheets: https://www.google.com/sheets/about/
- iCloud Numbers: https://icloud.com/#numbers

Besides having a sales scorecard, invest some time in creating a marketing scorecard, quarterly team scorecard, and a recruiting scorecard. Look into platforms such as Klipfolio or Monday.com.

Scoreboards can drive you to push yourself and break records to become the number one real estate agent in your territory, and dominate your marketplace. Get fired up; you are going to be unstoppable!

Take the Next Step
Execute Tactic 2.5: Create a Weekly Accountability Report.

PART III
ART

CHAPTER THREE

DISCOVER YOUR CORE VALUES

"I have learned that as long as I hold fast to my beliefs and values, and follow my moral compass, then the only expectations I need to live up to are my own."
—Michelle Obama

Do you find yourself asking, *"How do I get customers?" "What distinctive values do I offer my clients?"* Exactly! That is why I have set aside an entire chapter solely on values. The process of attracting prospects, capturing leads, and cultivating relationships into loyal customers starts with you having unique benefits to solve your clients' needs.

Without a doubt, every real estate firm can offer the standard real estate services expected by clients. However, what distinguishes you from everyone else licensed to buy and sell real estate in the same market as you, is "You."

As individuals, we are all unique. We all bring our combination of knowledge and resources to everything we do. In real estate, it is no different! What better place to identify the unique values we offer than our core values? They are our belief system that defines who we are and how we live our lives.

Discover Your Value Proposition

"What do you do for a living?" This question is by far, one of the most commonly asked when meeting new people. It is the most crucial

question that opens doors for new business opportunities, yet it is the most underused chance to shine. Many opportunities go missed when it comes to answering this common, recurrent question. The reason: lack of clarity in expressing our original and core values. Write down your answer to this question as you currently answer to others, and we will revisit it at the end of this chapter.

What we have done in the past years, what we currently do every day, and what we will be doing for many years to come is the reason why we exist. So when asked what you do for a living, the real question is, *"What is your purpose on earth?"*

You will begin feeling success the moment you visualize your purpose, and the astounding values you bring your clients and the world. Having clarity in your vision and awareness of your core values will bring you instant gratification even before you start marching your trek. Foreseeing how you can improve your customers' lives will change the path of your business journey and your life.

Develop the components of your real estate brand so you can better position yourself and attract people to want to do business with you. Lead the course of your business to a life-lasting success by clearly defining the core values of who you are, what you do best, and how you do it differently.

Unique Value Proposition

A Unique Value Proposition (UVP) is also known as a Unique Selling Proposition or Unique Selling Point (USP). While commonly referred to as an "elevator pitch" or an "elevator speech," I would argue that a unique value proposition is an elaborated elevator pitch, and a positioning statement that answers the questions: *"What do you do?"* and *"What makes you different?"* or *"How do you do what you do differently?"*

By definition, a unique selling proposition is a marketing concept that was first proposed as a theory to explain a pattern in successful advertising campaigns of the early 1940s. The USP states that such engagements made innovative propositions to customers that convinced them to switch brands. The term had developed by television advertising pioneer Rosser Reeves of Ted Bates & Company.

A unique value proposition is a promise of value that will be delivered, and a belief from a target market to experience proposed benefit. It is a concise positioning statement that explains the unparalleled benefits of a specific product or service. It is simple, short, and transparent.

Your UVP must resonate with a particular market. It must describe who you are, what you do, and what distinguishes you from the competition. Its purpose is to exhibit what solutions you offer that benefit potential customers in solving their needs, wants, and desires.

The UVP is merely the message that communicates your exceptional values to your target audience and your referral sources. It is created to strategically answer the most commonly asked business questions in the following categories:

VALUES

- *"What values do you offer?"*
- *"What is so unusual about those values?"*
- *"What are the benefits of those values?"*

PRODUCTS

- *"What products do you sell?"*
- *"What makes your product different?"*
- *"Why should people buy your product?"*

SERVICES

- *"What do you do for a living?"*
- *"What makes you different?"*
- *"Why should people choose to work with you?"*

As you can see, your unique value proposition is more about the gains you bring your customers than you. First, you must pinpoint and understand with whom you are speaking, because the benefits in your real estate services offered to sellers are different from those provided to buyers. Hence, your value proposition elements vary based on your target audience or customer.

Value Proposition Elements

To construct a compelling, unique value proposition, you need to explore five value characteristics: *customer, challenge, competition, compound*, and *close*. Those are the 5 C's to create a captivating UVP. I will refer to them as *"The 5 C's to UVP"* method.

Customer

A customer is also known as a target market or a target audience. Customers are a group of people—or businesses—tied together by wants, needs, and desires. They seek captivating solutions that cater to their most urgent needs. Your customer base is an audience who needs a supply that targets their demand explicitly. Hence, your target audience is anyone who needs the single most significant value you offer.

When product engineers package effective products and services, they evaluate the benefits every solution provides to each particular audience. That is why the more products or services you possess, the more audience you'll need to target. Therefore, you end up with a target audience for each product offering.

For example, if you say I provide real estate services to anyone and everyone, you are not targeting a particular audience. Hence, there is no apparent solution in said services. However, if you specialize and provide valuable benefits to home buyers, your target audience is home buyers. Similarly, if you provide solutions to sellers, your target audience is sellers.

Furthermore, there are different types of target audiences. A target audience doesn't solely have to be potential customers. A referral source, for instance, a business partner or an affiliate can each be considered a target audience because they can indirectly generate more customers for your business.

To maximize your customers and gain better insight into each target audience, consider creating a buyer persona. A buyer persona, or an audience persona, is a semi-fictional representation of your ideal client based on careful market research, past experiences, and insights about your existing client base. When creating each persona, think from the audience's perspective. Try to understand their goals, objectives, needs, and challenges so you can see what motivates them. Consider including their demographics, interests, hobbies, and day to day behaviors to find out what resonates with them. The more detailed you are about each persona, the better you'll connect with each audience on both the emotional level and the intellectual level.

> *Take the Next Step*
> *Execute Tactic 3.0: Determine Your Audience*

Define a Niche

Creating your niche is extremely critical to your real estate success. Targeting a specific client persona makes you an expert in that particular niche. Just like Allan Dib said in Success Wise, *"The Riches Are In The Niches."* Anyone can be licensed to sell real estate. Without a specific niche, you are a generalist with high competition. You'll have to spread your marketing budget and sales efforts so thin to capture a broad market and all kinds of homes and clients. Generalizing is one of the biggest reasons why most real estate agents fail, especially those who are new to the industry. What makes this even worse—most don't even realize it.

Your niche will help define you, but it doesn't mean you won't attract other clients. Agents may think that by targeting a specific niche, they will be losing potential opportunities. In reality, the opposite is true. By specializing in one market and one target audience, you will end up capturing other clients outside your niche. Most agents who try to relate to everybody won't attract anybody.

Spreading yourself too thin can waste your time, money, and energy. If you are not a large real estate firm servicing multiple territories with offices and teams across different areas, do not broaden your services. Working with buyers and sellers, buying and selling both residential and commercial is too general. When I ask most real estate agents whether they work with buyers or sellers, they answer "both." Such an answer presents two indications; you are either new or doing real estate as a side job. Top agents choose to work either with sellers or buyers, and most do specialize in working with sellers only. If I ask, "Which area do you serve?" Often, the answer is "Everywhere!" That is another red flag that this is a mediocre real estate agent who is desperate for new clients. Those two questions alone disqualify the majority of agents from getting hired. You don't want to be the average agent that specializes in everything to get a real estate deal—no matter how tough the market.

To build your niche, you need to be as specific as possible. Be neighborhood explicit, not city or county-specific! You may want to think of one kind of home, such as residential, rental, or vacation home. Work with one particular client type such as FSBOs only or Probates only. For example, do people in your area search Google for "Probate real estate agent?" What specific agents are they searching? Think of certain demographics or cultures in which you can connect. Is there a Spanish, Vietnamese, Chinese, German, Italian, French, or Arabic speaking community in your neighborhood? While being specific may seem counter-intuitive, with a smaller, and tighter net, you'll catch a lot more fish! Start small, and as you begin to dominate your unique niche, then you can expand.

There are enormous advantages to being specific in your real estate services. For starters, you would speak the language that most agents probably won't. You would know the culture and instantly build a rapport where other agents may not even have the chance. You would have the opportunity to connect with wealthy people outside of the United States. You would build stronger relationships where you can offer personalized services. For instance, pick clients up at the airport, book their hotel rooms, and show them around town. If you speak a different language, you can translate escrow documents and have experience in foreign exchange rates. If these examples don't apply to you, look for a specialized market to which no one is paying much attention.

Moreover, in a community of mainly residential homes, let's say about 5% of the real estate is farms and land. In this case, you will target people who own farms and live on acres of rural land. Although 5% is a small percentage of the total population, you will become one of the top agents in the area because you are focused on one specific niche. By specializing and fulfilling the necessities of a particular type of audience and property, you will know all about wells, septic tanks, and easements specific to the agricultural property. You will also be familiar with the different rules for square footage in outbuildings besides, whether they can have animals and the type of animals they can own. Choosing such a niche will give you a considerable advantage over other agents in the area. They would neither have the experience nor the knowledge you would have acquired over the years working in that specific niche.

Don't sound like everyone, so others don't overlook you. Make sure people know your expertise. Find out ways to serve a specific audience, be different, and be unique in your distinct niche.

> *Take the Next Step*
>
> *Execute Tactic 3.1: Create a Persona Worksheet to Gain Audience Insight and Determine Your Niche*

Challenge

A common mistake made by most businesses, including real estate professionals, is creating a one-solution-fits-all. Creating one product or one value proposition to target multiple audiences will lose its effectiveness. Each target audience will get lost once they come across your product and your brand because your message is too broad.

The primary goal of writing a value proposition is to identify how you and your services solve a specific customer's problems. However, the challenge comes in three different forms: a need, a want, or a desire— with the "need" being the most essential. That is why value propositions focus mostly on needs rather than wishes and desires.

To write a compelling value proposition, build valuable products, and properly position your brand, you will need to address the challenges of each target audience separately. Each solution or product provides different features and benefits to solve various problems for each target audience.

To be productive in solving your customers' needs, first identify your audience, then look for their challenges—needs, wants, and desires. For example, a seller's needs and desires are different from those of a buyer. When you're representing sellers, your message and your brand have to speak to sellers. Therefore, your value proposition must resonate with either a seller or a buyer.

Use *The Three Levels of Customers Demands* to determine customers' problems and the urgency level of each challenge.

1. Need: an urgent necessity with a *must-have* solution to a problem.
2. Want: an essential demand with a *should-have* solution to a problem.
3. Desire: a crave or a wish with a *could-have* or *nice-to-have* solution to a problem.

Understand the difference between all three challenge levels for each target audience. Address each concern separately starting with their urgent needs first. Solving clients' problems is imperative in making you sound more relevant and useful in your communication. This prioritization method is known as the *MoSCow* Method.

While serving a need is an absolute requirement, fulfilling a desire is far more rewarding. If you foresee all obstacles that stand between you and your customer's wishes, you might be your customer's hero. For instance, suppose Sally, the seller **needs** to sell her house. She **wants** to sell it for $500,000 but **desires** $550,000. The fair market value maybe $500,000. As an agent, what do you do?

- To fulfill the *need,* you must sell the house.
- To fulfill the *want,* you must sell it at a minimum of $500,000.
- To fulfill the *desire,* suggest home upgrades or renovations of $10,000, so you could list at $560,000 and sell at $550,000.

....

: **Take the Next Step**

: *Execute Tactic 3.2: Discern Audience Challenges*

....

Competition

Most people are aware of the high level of competition within the real estate industry. What people aren't fully aware of is that the values each real estate office and each real estate individual brings to the table. Customers are equally frustrated to choose the right agent to represent them, as agents try to get signed contracts. My advice would be to focus far more on the values that you bring your customers rather than focusing on your competition. More importantly, focus on how to communicate those values to your marketplace better.

Besides, your fellow real estate professionals in your office are your competitors as much as the agents working for other firms in your town. Instead of avoiding your competition, try talking to them. You would be surprised at how helpful people are. Find out what's working for them and their clients. Ask about the value elements that worked best for them. Also, inquire about what is working with their current clients. For example, "Were there any guarantees?" "Did specific marketing packages have better success?" "What worked most in listing presentations?" Don't just copy what they did or what they currently do. Observe, learn, and help your clients in your unique way. Be different, but, of course, better than your competition.

> **Take the Next Step**
> *Execute Tactic 3.3: Spy on Your Competition*

Compound / Product

The real estate industry is a relationship industry. It is personal! Real estate professionals do not sell goods, merchandise, software, or platforms; neither do they sell services. People hire real estate professionals for their keen knowledge in the housing industry. People seek agents, so they make the right decision with their most significant asset or investment in their lives. The real estate professional is the product. "You" are the product. You, your knowledge, your dedication, your commitment, your values, and your real estate brand are what matters to people. You are the #1 reason people hire you.

When you are the product, you become the brand—your own, personal brand. Your brand identity is what determines how you are known and what you are known for in the world. You control the values you bring and the benefits you want to offer. You decide whom you want to serve and what you help them achieve. Real estate is not a job or a career—it is your purpose, and you have full control over it.

You are much more than "a real estate professional." You are the hero that positively changes home buyers and sellers' lives for the better. Know your *purpose*, find your *passion,* and showcase your *proficiency*. Create your *proverb* or tagline. Say it out loud, and become the *product* people enjoy to have in their lives. You are not just a real estate agent. You are a combination of a purpose, a passion, and a mission to serve people's needs, wants, and desires.

> **Note**
>
> *I named this section compound to demonstrate that the product is a blend of who you are as a person and the services you offer as a company. Your customers don't just look for "real estate services" when deciding who to hire. Their best valued real estate product is a mixture of "You" and the services you provide. The language used in constructing a Unique Value Proposition is also called a Language-Market Fit (LMF).*

A *Language-Market Fit* is the effectiveness of your brand message and how well your language resonates with the people you're trying to reach. There must be a mutual correlation when speaking to clients in-person, the words used on your website, and sentences utilized in marketing. The language you use is the bridge between your customer's problems and the solutions you offer. Your style will fit your market to help drive your success. You can reach new people, retain customers, scale growth, and move nations with the language you use. Hence, "*I Have a Dream,*" by Martin Luther King Jr.

Close

While reading through this book, you'll notice that I cover every foundational concept from choosing your target market to solving their needs. Then, I show you how to apply these principles through business development, marketing, and sales. Ultimately, I tie it all together so you can have a thriving real estate career and dominate your marketplace. In close, two questions will spur:

1. *Have I clearly explained all of the benefits that I can offer my real estate professionals?*
2. *Have I painted a concise mental image of what their lives would look like after I've solved their needs?*

Similarly, before closing your clients, ask yourself these two questions. By doing so, you will win every customer if you confidently answer, "Yes." Test whether you have reached success by asking your target audience to answer related questions. Can they envision their lives after you've rendered real estate services? Will they be delighted with the results? Will they use you again? Will they refer you to their family and friends? Such questions will take you over the top and open conversations you may not have been able to have before. If everything checks out, go ahead and ask them to take action. Schedule that appointment, and your customers will be delighted to sign the contract.

Other forms of closing calls-to-action early in the process can be a phone call appointment or a face-to-face meeting. Others such as visit our site, like our page, and follow us are used to progress leads through the closing process. Nurturing client relationships will be covered in upcoming chapters. However, the purpose of this section is to demonstrate that you should only close customers once you are sure your values have been received and understood.

Why is Value so Significant?

To create the clear one-liner that clarifies who you are, what you do, and how you do it differently. To produce a single statement that defines whom you do it for, and how you make people's lives better. There must be a heavy emphasis on value. Your authoritative and authentic value statement needs to highlight all of the compelling benefits you provide.

Usually, the main focus of your customers is on value. In most business situations, clients rely on their decision-making entirely on the value provided. Whether a buyer is purchasing a product or hiring for service, shoppers usually evaluate whether the product value is worth the price tag, or service provided is worth the associated cost. Regardless, customers measure whether they want to buy or hire based on **benefits** versus **costs**. They buy when benefits outweigh the costs. So, there are only two components to consider when measuring your product values:

- The benefits your services create for your customers
- Their fees in terms of price, plus any ancillary costs of ownership or usage. For example, how much time do they have to devote to buying or using your product or service?

The value a customer receives is equal to the benefits of a product minus its costs. (*Value = Benefits - Costs*). The more you increase the benefit factor to your customers, the more likely you will win because your *value* will be far higher than those of your competition.

That said, escalate your perception of value by either increasing your benefits or decreasing your costs. You can charge the highest commission when your benefits to your customers exceed the total costs. Your customer's perceived value has to surpass your commission and all other costs associated with the real estate transaction. In other words, your customer's evaluation of their gains must be higher than all expenses. This concept is highly critical in the negotiation process but often misunderstood by most agents. How are you justifying your commission with the benefits you offer your clients?

> **Take the Next Step**
> *Execute Tactic 3.4: Measure Your Current Benefits*

"You are as great as your unique values."
—James Tyler

Let's dig deeper into the unique value proposition approach and explore its two most essential elements: uniqueness and value. Uniqueness will help you analyze how you can increase your credibility, add more benefits to your customers, and better position yourself.

Discover Your Uniqueness

Uniqueness is a physical characteristic that identifies a perceived scarcity derived from a specific value. This value can either be a remarkable quality of an agent or an exceptional service provided by a brokerage firm. Uniqueness exists when you have distinct attributes that are not offered by others in the marketplace.

How Are You Different?

Whether you are a real estate agent or a broker of a firm, there must be a bizarre quality about you or your business. Novel elements differentiate you to stand out from your competitors and elevate the values you bring your customers.

You can point out differentiating factors such as leadership, accountability, experience, flexibility, and accessibility in your real estate brand. Also, you can highlight your firm's reliability in getting things done, and the speed in closing the home buying or selling process. Besides, the quality of your listing presentation and the effectiveness of your marketing package could have elements of uniqueness. However, whatever distinguishing characteristics you elevate, they must be in the customer's best interest. They shouldn't be about you, or your real estate firm. Customers are interested in knowing what is in it for them, and how you are going to save them time and money! For example, if you keep up with the latest real estate news, you would deliberately emphasize how your real estate brand is trending and reliable.

Consistently and intentionally underline particular qualities that you or your business has and communicate such characteristics to your marketplace to rise above the competition. For example, instead of saying, *"We provide the best customer service,"* you could highlight a valuable team, *"Communication Guarantee."* A guarantee would separate you from your competition by promising your customers to stay in touch regularly. It gives the client the power to terminate the contract if the agent doesn't follow through. The communication guarantee is especially crucial to sellers with expired listings who complain about the lack of communication from the previous real estate agent or office.

> ### Take the Next Step
> *Execute Tactic 3.5: Create a Communication Guarantee Policy*

Communication Guarantee

"I will keep communicating with you every week to give you a progress report on the sale of your property. I will speak with you within 24 hours after each home showing. If I fail to do the above and not honor my Communication Guarantee, I will give you $100 cash each time up to $500*.

*Provided that there are no immediate personal urgencies or situations such as sickness or death that I need to attend."

If you are a real estate agent who wants to promote your real estate business as a personal brand, you will become the product. By default, you will mainly focus on building a name for yourself. Elevating your brand doesn't mean the perceived value is self-promotion. Whatever motto you pick to make your hook, make sure it is about your customer, and not you. For example, striving to be known for *"the top real estate producer"* in your marketplace does not show any direct benefit to your customer. However, offering a *Risk-Free Guarantee* is beneficial to your customer, and holds you to the standard of keeping a high level of customer service.

Providing your buyers, sellers, For Sale by Owners, and Expireds a secure *exit listing guarantee risk-free* would separate you from the rest of the agents in your space. You can then overcome the contractual hiring objection by promising a 30-day rescission policy within your *Risk-Free Guarantee:*

"If you feel like we are NOT a good fit within 30 days; you have the right to cancel."

This exit listing cancellation clause allows customers to fire you if they are displeased with their experience or you fail to meet the requirements of the agreement. People don't like to become locked in a contract, and this takes the risk away from their decision to not hire you.

....
Take the Next Step
Execute Tactic 3.6: Create a Cancellation Guarantee Policy
....

No-Risk Guarantee or Cancellation Guarantee

- *"I guarantee that you have the right to cancel this Listing Agreement with NO COST or obligation to you within 30 days if you are not happy with my services*."*
- *"We are so confident that you will be more than happy with our real estate services. We guarantee you the right to cancel at any time before accepting an offer to purchase your home, with no penalties or obligations. You have the right to cancel our listing agreement should you feel our service doesn't live up to our promise to you*."*
- *"We offer an easy exit listing where you have the right to cancel the Listing Agreement at any time with no obligations."*

You may also add: *"*I only ask you to allow me the opportunity to correct any difficulties within 48 hours. If you are still unhappy, you have the power to exit the Exclusive Agreement with no hassles. It's that easy."*

Your unique values don't have to be qualities that you possess entirely right now. They could be benefits that you or your business improves upon and develops over time, for example, the speed of finding the right home for your buyer or processing the sale of a house. In this case, you may want to develop a *Close on Time Guarantee* policy.

Your unique values must deliver on promises made, and they must be about your customers. Uniqueness must be about the "values" and the "benefits" you bring your clients. Whatever distinguishes you from your competition and makes you different have to matter to the customer. For example, if it takes you or your real estate office 45 days on average to close a sale, separate yourself from your competitors by creating a *Performance Guarantee* policy. This way, you can overcome the seller's fear of hiring you and not selling the house objection. *"What if you don't sell my house? What happens?"* You can promise something like:

- *"I will sell your home, or you don't have to pay me."*
- *"I will sell your home, or I will pay you $500."*

Alternatively, you can combine both the *Close on Time Guarantee* with the *Performance Guarantee* into something like this:

- *"I will sell your house for free if I don't sell it within 45 days."*

> **Take the Next Step**
> *Execute Tactic 3.7: Create a Close Guarantee Policy*

Performance Guarantee or Close Guarantee

- *"I guarantee performance in closing your home within 45 days; otherwise, you may exercise your Cancellation Guarantee policy without any obligations."*
- *"If you list with us, and you sell it yourself, you pay us nothing!"*
- *"Guaranteed sold, or we will sell it for free."*

Regardless of whether you are a new agent, a top producer or specialize in selling luxury homes, your hook doesn't have to be who you are right now. It can certainly be what you want to be known for in your marketplace, as long as you deliver on your promise. Document who you are at the moment, speak the truth and strive for what you eventually want to be. Let people learn who you are, and let them watch you develop into whom you want to become. If you're going to be known as *"Rose, The Real Estate Go-Getter,"* then **push, progress** and **produce**! Eventually, you'll become the agent who hustles. Your *hook* is *"you hustle for your customers."* If you want to be known for *"Rose, The Aggressive Real Estate Negotiator,"* then negotiate well for your customers. Get them to brag about all of the monies you've made them or saved them.

> **Take the Next Step**
> *Execute Tactic 3.8: Create a Marketing Guarantee Policy*

Marketing Guarantee

"We give you a written, 75-point marketing plan when we list your house, which outlines what we will do to get your house sold within 45 days and for the most money. That way, you are certain of what is being done to market your house, and there are no surprises. If at any time we do not live up to our

agreement, you can cancel your listing with no penalty."

You will have the best competitive edge, the highest list-to-sale-price ratio in the neighborhood, help price the property correctly, and put conditions in place for a smooth sale. You can implement a *Certified Pre-Owned Home* (CPOH) Program as people want transparency from the home buying and selling process in today's world. CPOH makes that process a cleaner, more efficient, and more trusted model. Buying and selling a home should be a positive experience for both buyers and sellers.

> **Take the Next Step**
> *Execute Tactic 3.9: Create a Certified Pre-Owned Home (CPOH) Guarantee Policy*

Whatever you want to be known for or however you want to brand yourself, make sure you have a mark. You are unique, you are unstoppable, and you are a go-getter. Create special programs for your clients. Have fun branding yourself and your business with uniqueness. It is powerful, and it works—as long as you're honest and genuine!

Elevate your values by thinking of other creative guarantees that will make you unique and separate you from your competitors:

1. Communication Guarantee
2. Seller's Guarantee Program, No-Risk Guarantee or Cancellation Guarantee
3. Performance Guarantee or Close Guarantee
4. The Extensive 75-Point Marketing Plan
5. Certified Pre-Owned Home Program
6. The High-Quality Marketing Materials Guarantee
7. The Social Media and the Internet Guarantee Program
8. Engaging the Real Estate Community Guarantee
9. Traditional Marketing on Steroids Guarantee
10. Merge all of them into one complete "Marketing Guarantee Program."

> **Take the Next Step**
> *Execute Tactic 3.10: Discover Your Uniqueness - Construct Your Three Unique Must-Have Values*

Your Unique Value Invitation (UVI)

Unique value invitation is a formal and polite invitation to demonstrate the top three unique values you offer your marketplace. It is a message to those who show some interest in what you do. Its function is to convert invitations into appointments by expressing your benefits and uniqueness. The request could be conducted by phone or online to convince, persuade, and schedule in-person meetings.

Think of a new digital platform that you need from a software-as-a-service (or SaaS) company. For example, you research a few Contact Relationship Management (CRM) vendors. You read their websites to understand all of the features, benefits, and price points. After you compare the top two or three vendors, you feel the need to see a demo before committing to one. The purpose of the demo is to find out which platform would be the best fit. After all, you are looking for the CRM that solves your needs. You benefit from the demo because you get a sneak peek into what you would be dealing with after you sign up. Vendors also win by getting the chance to proving themselves after comprehending your needs. They will be revealing their outstanding service, benefits, advantages, social proof, and client reviews.

A Unique Value Invitation is no different. It is your opportunity to show that you are the best agent for the job. It is your chance to prove yourself to those who aren't ready to commit yet. All you need is fifteen minutes of their time.

Agents often come across people who aren't ready to meet with them or hire them just yet. Generally, these are cold leads that usually want to find out a little more information. Instead of rushing the closing process or risk losing them by asking to meet in-person, you should allow them to learn more about you, and the benefits you provide.

Politely invite your prospects to a phone conversation or a web meeting. Find out what their demands and pain points are. Explain the top three unique benefits that you can offer them to solve their needs! Perhaps you want to implement and share the guarantees mentioned above in your real estate business model as your top qualities to reveal during your Unique Value Invitation (UVI).

Finding out whether you are a good fit, and whether you can solve your potential customer's needs is imperative. You save time and pitfalls when you determine if there is a fit. When you invite your target audience to understand them and also explain your values, it shows that you care. Prospects won't perceive as another pushy agent. It removes the sales pitch factor and the sense of being a selfish real estate salesperson. It makes you highly credible since you become solely focused on the customer's challenges and your values that solve those needs.

Use the Unique Value Invitation (UVI) approach as a must-have since it will always be appealing to your target market. It acts as the gateway to solid appointments. Utilize it as one of your most significant opportunities to connect with your target audience. UVIs are the bridges that move prospects from cold leads to potential customers. Candidates will be ready to make decisions and meet with you face-to-face, only when convinced that your values will offer them significant benefits. Unique value invitations will do just that. They often lead to appointments and eventually loyal customers. I urge you to master them and make them part of your real estate business model.

> **Take the Next Step**
>
> *Execute Tactic 3.11: Test Your Unique Value Invitation*

> *"I am rare, and there is value in all rarity; therefore, I am valuable."*
> —Og Mandino

Discover Your Values

What benefits do you bring to your customers? If you master the art of increasing the recognition of value, domination will be inevitable. The most effective marketing and sales strategies apply the tactic of improving the viewpoint held by the target market regarding the value of the products or services offered. The more you increase clients' understanding of value, the more they will likely work with you, and of course, pay your asking commission without any objections.

Value, by definition, is the number of goods or benefits in services offered in exchange for something else. For anything to have value, five specific characteristics must be present and perceived by the user or consumer. The more you master elevating each aspect of these five elements, the higher the value. These five value characteristics can be used as methods to increase the perception of the importance of any product, not just your brand or your real estate services:

1. Demand
2. Utility
3. Scarcity
4. Flexibility
5. Credibility

Demand

It is the need or desire for a specific good or service. It is an indispensable component in creating value. Without demand, any amount of supply is meaningless. Sometimes though, it can be formed by merely lowering the price. This method is detrimental as opposed to increasing the desire to increase demand, therefore perceived value. For example, reducing your commission to create a need puts you at a disadvantage. Instead, think of ways to elevate the desire of customers wanting to work with you so you can create a constant demand for your services without lowering your price.

On the other hand, prices may rise too high even when there appears to be keen interest. High rates cause some people to look for alternative solutions, known as *The Principle of Substitution*. People seek alternative solutions when perceived value is less than the asking price, or the rate is much higher than the marketplace. Ensure a balance between value and cost, even if there is a pressing need. This way, your customers don't seek to work with other competing agents.

One final element of demand to consider is a person's ability to pay for an item. *Effective Demand*, the level of the market that represents a real intention to purchase by people with the means to pay. Purchase ability means a prospective buyer has enough disposable income available to satisfy her or his needs or desires. A person may want a million-dollar home, but if she or he can't afford to buy it, then that person's demand doesn't count. That is why most real estate professionals work with people that have sufficient demand (e.g., pre-qualified buyers).

> **Take the Next Step**
> *Execute Tactic 3.12: Discover Your Values - Demand*

Utility

It is the ability of a good or service to satisfy human demands (needs, wants, or desires). It is the clarity and degree of usefulness to prospective users. For example, a potential buyer may perceive a listed house of sale as useful only if it has enough bedrooms.

Furthermore, if there is no perceived use of something, then there wouldn't be any sense of value. The value will lose its usefulness if there is a limitation, restriction, or a lack of clarity. For example, Government restrictions on the size and placement of buildings may mean that a particular type of structure is prohibited. Public or private limits on land can affect its value. Any restrictions can impact the utility, and consequently, the value of the property.

Increase the qualities of being helpful by clearly defining your services and the characteristics that are valuable, beneficial, functional, and essential to your customers. Look for the advantages that you have over your competitors. For example, being accessible 24/7 is an edge that your competitor may not have. You speed up the closing process when you provide potential sellers with a step-by-step guide to selling during a listing presentation.

Find ways to be useful in creating functional advantages that can provide several benefits. Think of the different types of benefits you can offer your customers. There are financial, emotional, physical, and spiritual benefits behind every single product offering.

> *Take the Next Step*
> *Execute Tactic 3.13: Discover Your Values - Utility*

Scarcity

Scarcity is the perceived supply of a good or service relative to the demand for the item. It is the level of exclusivity of your offer. If there's an unlimited supply of something, then it's perceived to have little value. Of course, the rare item must also be useful (have utility). Scarcity is most effective when usefulness is discerned and recognized. "Scarcity" and "Utility" must both be present. Things that are scarce but not useful have little value (e.g., common fossils). Just like entities have little value if they happen to be plentiful, even if useful (e.g., air or an abundant source of water).

People generally perceive land to be valuable because there's a limited supply of it. This notion of inadequacy feeds the anticipation of real estate buyers who feel they're buying a property as an investment that will increase in value. Value derives from scarcity due to uniqueness; for example, the fact that there's only one of a particular house in a given location. If people desire a distinct home in a specific area, then more value may be created.

Increase the perceived value of a good or service by creating a limited supply of the gains offered to each segment in your marketplace. Combine the element of scarce with uniqueness and versatility to create an exclusive effective demand. Hence, you create desire and attract the most qualified buyers and sellers by elevating your values through uniqueness, usefulness, and rareness.

Organizations usually create scarcity through discounts and rewards, offering cash refunds, seasonal rebates, and time-sensitive coupons. Since discounts and rebates do not apply in the real estate industry, how would you use this principle? You can leverage scarcity by being innovative in increasing your perceived values. For instance, if you are the selling agent, show the listing agent that you have limited buyers who are qualified and interested in the seller's particular house.

Furthermore, if you want to increase the perceived value of who you are and what you do, create a scarcity that isn't offered by any other agent in your area. For example, you may want to become an agent who has a diverse number of buyers. You can do so by sending a quick survey to your existing database with these two questions:

"What does your next ideal home look like?"

"What is your favorite city that you want to live in?"

You can get creative with your questions to profile your existing relationships. Being innovative will give you an advantage over the other agents in your area. You will have viable reasons to connect with these potential buyers should their ideal homes be on the market. More importantly, you have just created scarcity to those sellers.

Likewise, send another one-question survey to existing homeowners:

"When do you think you would be selling your house?"

    A. This year
    B. Next year
    C. In 2 years
    D. In 3-5 years

You now have potential sellers to reach out to when the time is right, and when they are ready to make a move.

Besides creating more surveys, you can come up with new and creative ways to be unique and useful. For instance, offer one-on-one exclusive consultations to past customers and referrals only. This way, you can leverage your existing clients to send you ready-to-commit referrals. Make yourself be the singular real estate agent that shines in every buyer and seller's eyes. There are plenty of real estate agents in the marketplace, but there aren't many who are unique with differentiating rare qualities. Similarly, there aren't many who provide substantial benefits consistently. Create scarcity and inevitable demand by being a unique and useful agent. Increase your real estate values by creating exclusive offers to limited audience only.

: **Take the Next Step**
: *Execute Tactic 3.14: Discover Your Values - Scarcity*

Flexibility

Flexibility may also be known as transferability, or the ability of a good or service to satisfy human demand (needs, wants, or desires) even when the supplier isn't present. If a supplier is absent, the benefits of the services provided must be flexible and transferable to an agent who can serve customers as anticipated without any limitations or conditions. Nonexistence of flexibility might restrict benefits from being supplied to customers as expected and therefore lessen the value of the service.

As a real estate agent, impose no restrictions on flexibility and transferability. Ensure clients that you have a support team behind you that is true to the qualities and values you promise. In case you're not available, your customers will receive the same great extraordinary benefits as if you were still there.

: **Take the Next Step**
: *Execute Tactic 3.15: Discover Your Values - Flexibility*

Credibility

Credibility defines the attitude of a receiver which references the degree to which a source is seen to be believable and trustworthy. Online reliability has become an important topic since the mid-1990s because the web has increasingly become an information resource. Like every other industry, real estate has also moved prominently online. Hence, to increase your online credibility as a real estate agent, you have to address its three main components: *competence, character,* and *care*. These three dimensions play an essential role in your online credibility.

Competence concentrates on expertise, skill, or knowledge in a subject matter. While character refers to the goodness (i.e., honesty, or trustworthiness), caring focuses on showing concern and empathy. Although an individual may exhibit one or two of these qualities, the best and most respected exude all three. Therefore, to become credible, it would be best to show confidence in your expertise and empathy towards your customers.

While caring includes subjective components such as sources of dynamism, charisma, and physical attractiveness; Competence and character have objective and subjective elements. Skill can be subjectively perceived, but also includes relatively actual objective characteristics of the source or message (e.g., credentials, certification, or information quality). Similarly, trustworthiness is based on subjective factors, but can also include objective measurements such as established reliability when customers feel their needs are satisfied.

Start building credibility with your customers today by following the components of competence, character, and care. To apply each part, let's visit each in more detail.

Competence

What are you doing today to show your customers that you have the skills and knowledge necessary to satisfy their needs? Create a blog to share valuable real estate information with your potential buyers and sellers. Take the content from the blog and share it on social media. Re-purpose such content for YouTube, Facebook Live, or Instagram TV (IGTV). Also, you can host monthly events at local meetups to educate your community. Earning the "I am a real estate expert" reputation doesn't happen overnight. You've got to work hard for the knowledge. Are you a lifelong student of your subject matter? Hire a real estate coach. Attend real estate conferences. Learn, practice, and teach your community so people can perceive you as an expert in the subject matter.

Character

How credible and reliable are your claims to the benefits you bring your customers? Trust is built gradually by many factors experienced by your clients. Some of these elements could be the authenticity of your real estate services, business atmosphere, proven evidence meeting customers' needs, and social aspects related to past clients. Nevertheless, there are only three ways to build trust using the character element:

1. Do what you say you're going to do.
2. Deliver what you say you're going to deliver.
3. Show you delivered what you said you were going to provide.

How trustworthy are your promises and the values you claim to offer? Even if customers aren't asking you, they are most likely thinking about it! Be accountable for your commitments. Be honest, proactive, and show up to your appointments on time. Be reliable with your actions. Be sincere, responsive, and deliver your services to the deadline you agreed. Ask customers for feedback, testimonials, and reviews. Engage with your customers on the web, social media, and in person. Look your best, dress to impress. Generate proven evidence of your trustworthiness. Build trust and meaningful relationships by allowing your business to be all about your customers.

Care

What do you do that shows how much you respect your customers and care about them? What do you do to show them that you have their best interest in mind? How do you positively influence them and their desired outcome?

Build Rapport

The first rule to building rapport is to ask genuine questions and let people talk about themselves so they can relax and "open up." If you are meeting in person, check your appearance as first impressions do count. Identify common ground and use small talk to learn more about your customers and their shared interests. Use the mirror and match method at all times; listen carefully and attentively. Listen to their voice and dig deep into their body language. Watch your audience's body movements, including gesture, posture, and expression. Match their speech patterns, such as tone, tempo, and volume. Modulate your style, pace, and cadence to match your audience. Sync your behaviors to connect and build rapport.

If you're speaking on the phone, your tone is what matters most. Although body language accounts for 55% of human communication, you don't have the advantages of the physical presence when you are on the phone; you have tone and words. Your words account for only 7% of communication, and your tone of voice accounts for 38% of communication. Therefore, the words of your scripts don't matter as much as the way you say them. Practicing and mastering your tone of voice is the most critical aspect of persuading others in building rapport and trusting to work with you.

Positive Influence or Change

Mirror and match positivity. For instance, if you have an upset buyer, you obviously won't mirror and match their frustration. In such a case, you must lead and influence with a positive mindset to make a positive change to build trust. Be humble and honest, regardless of your customer's situation. Never argue, complain, or blame. Stay optimistic and have fun with your customer. Build credibility and excellent reputation by looking for their best interest. Show that you care by being present in the moment at all times.

Selling and buying real estate is somewhat intimate. Get a bit personal. Offer positive recommendations with multiple options based on clients' intentions and emotions. Understand their motives and take immediate actions that will result in concrete solutions and positive change in people's lives.

Emotional Intelligence

Be empathetic, not just sympathetic. Listen to your customers, really listen, and try to anticipate their needs through their emotions. Try to understand how they feel; tap into their feelings! Building rapport alone isn't enough to show that you care. You need to understand emotional intelligence and use it to benefit your customers.

Show respect by asking the right questions to not only find out their motives but also to understand their emotional stage before you offer to help. For example, instead of the sensitive, sympathetic method: *"I know how you feel,"* use the empathetic approach, *"I've purchased a home before, and I know exactly how you feel right now. It is normal to feel anxious and nervous. Here is how I can help make the situation better."*

Take the Next Step
Execute Tactic 3.16: Discover Your Values - Credibility

Calculate Your Value Score

Let me introduce you to the "Value Score" system now that you have explored all of the components that make up the values to elevate your real estate business.

A value score is a statistical measure to determine the uniqueness of each of the value components: Demand, Utility, Scarcity, Flexibility, and Credibility. It measures your favorable resources against your competitors. When you compare your values against those offered by your competitors, you can determine which elements hold the most weight. Therefore, you will compete and win your customers on your top-ranked advantages. For each target audience, insert their challenges and the solutions offered to address each problem. Compare what you believe your value score is in comparison to your competitors. Then compare your values against your competitors for the five value components (Demand, Utility, Scarcity, Flexibility, and Credibility). Pick only the top four benefits under each one of the five value components, list the four benefits YOU offer your customers. Then, weigh each benefit on a scale from one to five, one being the least unique and valuable, and five being unique, in comparison to the real estate industry. For instance, suppose you had 11 Demand value elements on your worksheet. You will only choose the top four. On a scale of one to five, you would want to ultimately get each value element to be at five totaling one hundred *Value Points* (VP).

Higher value score indicates more benefits that you bring to your customers. Whether values are characteristics of you or your business; benefits must be as unique as possible and all about the customer.

Be unique and provide proof or evidence to substantiate your real estate values. Define your advantages by the exclusive benefits your customers gain, not by what you do for them.

Here's an exercise that might simplify this technical approach:

On a scale from one to five, write four benefits for each value component and an associated score for each resource.

I. DEMAND: *Benefits that create appeal and desire*

S A M P L E   B E N E F I T S

1. Knowledge of Real Estate Market	(1-5)
2. Years of Experience	(1-5)
3. Negotiation Skills	(1-5)
4. Marketing Advantages	(1-5)

Total DEMAND Score: *(4 to 20 Value Points)*

II. UTILITY: *Usefulness and clarity of benefits*

S A M P L E   B E N E F I T S

1. Marketing Packages	(1-5)
2. Process Guides and Brochures	(1-5)
3. Intuitive Online System & Private Facebook Group	(1-5)
4. Access to a Large Partner Network	(1-5)

Total UTILITY Score: *(4 to 20 Value Points)*

III. SCARCITY: *Limited availability of useful benefits*

S A M P L E   B E N E F I T S

1. Communication Guarantee	(1-5)
2. Happy Buyer & Seller Guarantee	(1-5)
3. We promise to treat you like a person, not a lead	(1-5)
4. Work with Professionals Guarantee	(1-5)

Total SCARCITY Score: *(4 to 20 Value Points)*

IV. FLEXIBILITY: *Flexible and transferable benefits*

S A M P L E   B E N E F I T S

1. Best Communication Guarantee	(1-5)
2. Available when convenient for you	(1-5)
3. Support team guarantee	(1-5)
4. Responsiveness Guarantee	(1-5)

Total FLEXIBILITY Score: *(4 to 20 Value Points)*

V. CREDIBILITY: *Believable and trustworthy benefits*

S A M P L E   B E N E F I T S

1. Online Reviews	(1-5)
2. Testimonials	(1-5)
3. Social Proof	(1-5)
4. Web Presence	(1-5)

Total CREDIBILITY Score: *(4 to 20 Value Points)*

[1] *Not unique* ----------------- [5] *Highly unique and beneficial*

Demand + Utility + Scarcity + Flexibility + Credibility = Total Value Score

Add them to determine your total value score. Is it between 80 and 100?

Your values increase the more you elevate any of the above five elements of benefits. For example, if you are competing with another agent to get a listing, negotiate based on a comparison of your value score against the competing agent. By comparing the benefits offered to your client, you will overcome the client's objections. Also, you will stand out from any other competing agent due to the tremendous benefits you bring to the table.

··
: *Take the Next Step*
: *Execute Tactic 3.17: Calculate Your Value Score*
··

∼❖∼

CHAPTER FOUR

Elevate the Power of "You"

"You never know how strong you are until being strong is the only choice you have."
—Bob Marley

Elevate the power of who you are and what you do to build a solid foundation for a long-lasting, successful business. Increase relevance, and connect with more customers by making your real estate business about "You," not just your business. There is more to success than an elevator speech and a unique value proposition.

Where most brands go wrong is after they launch their business, and open their doors, they jump directly into marketing. When, in fact, there is still plenty of foundation and ground to cover before telling the world about the top unique benefits they offer. People are initially more concerned about "who" is behind the business than "what" the company does. As Barbara Corcoran once said, *"People want to do business with someone they like. If people like you, they're going to want to do business with you. If they don't, you're going to have an almost insurmountable obstacle to overcome."* Hence, it would be best if you addressed "You" before your real estate business.

Who are you, and what's your story? How can you use your personality to connect with mass clients? How does your real estate business reflect who you are? To answer these questions, you must make and wear the persona in which you want people and clients to see.

For starters, you need to build a brand around you, your target markets, and their values. Your message must be clear and understood through various communication channels—whether offline, online, social, or mobile. After all, effective communication is what drives our professional and personal success. To have prosperous marketing and sales efforts, and a thriving real estate business, it is no different. Your brand message must be concise and useful. You and your message must resonate with people. You must build a reliable and trustworthy online presence, and your online identity must start with "You." Thus, if you want to dominate, you must communicate who you are, and the usefulness of your business dramatically. Does anyone other than yourself know who you are and what you do better than yourself?

Every real estate agent wants to rank on the first page of Google Search. However, when I researched real estate agents in my city, I was shocked to see only three websites with acceptable "About" pages in the top one-hundred results. This outcome meant that only 3% of agents in my city took the time to write about themselves. Of course, they were the prime agents that I am already familiar with and have seen their signs all over my neighborhood. This research saddened me. Is it possible that they don't realize the value of having a personal brand? Do they need to be guided to build a reputable online presence? Perhaps in this chapter, I can shed some light into a significant problem that most agents must assess; that is the fundamental essence of building a respectable online identity.

The top-producing agent ranked on the first page of Google Search, and the other two results were on page two. The rest were embarrassing. Most "About" pages were one or two paragraphs and seemed to be written by the web developer. Some didn't even have an "About" page. Others had only the agent's contact information. The majority of the other websites were the boilerplate company website with "Our Agents" page and contact information.

If you aren't one of the top three real estate agents in your area, how are people learning about you? Advertising on Zillow and Yelp is excellent, but that's what most agents do. How are you different? People want to connect with you on a personal level. They want to get to know you, like you and trust you before they sign that contract. If the "About" page on your website has nothing about you, how do you allow outsiders to connect with you?

The goal of this chapter is to move you from 97% of mediocre agents into the top 3%. "You" are going to write your tagline, elevator pitch, unique value proposition, storyline, biography, and your executive summary. When doing so, don't hire a third-party service to write your most potent brand messages that you are about to discover.

Craft Your Proposition

In the previous chapter, you learned about the elements that make up a unique value proposition. You identified your target audience and their challenges. You compared your values against your competition and elevated your core values. Now, what do you do? You build your core character—the person and the brand you see a decade from now.

Craft Your Tagline

Your tagline is the slogan that tells your target audience who you are. It resembles how you want others to perceive you and your business. Since "You" are the product as a real estate professional, you would be writing a slogan for a personal brand. (See "Compound" under "Value Proposition Elements" in the previous chapter.)

There are three pillars to elevating the power of "You" and writing a personal brand's tagline: *purpose, passion,* and *proficiency.* Use these three elements to construct your "Tagline." Start brainstorming your tagline by writing multiple variations. Once you have a few versions, summarize them into one. Evaluate whether you're satisfied with its concept, edit if you need to, and then finalize. To learn more about *"The 3 P's to Elevate the Power of You,"* please see the previous three chapters.

If I were to reactivate my real estate license and do real estate again, here are the steps that I would take to create my tagline:

1. **Purpose**: I feel the need to help, educate, and support myself and others.
2. **Passion**: I love to read, run, hike, snowboard, golf, and have fun in the sun.
3. **Proficiency**: I am proficient in real estate, mortgage, sales, marketing, programming, web design, and business management.

To **brainstorm**, I would write a minimum of three examples:

1. *"I help home buyers, and sellers benefit from my real estate sales, marketing, and technical expertise. I am educated, positive, outgoing, and fun to be around."*
2. *"You know how people in need of real estate services are always looking to save and make more money? Well, I help those people save money due to my extensive knowledge and usefulness in real estate and*

technology."

3. *"You know how buying and selling real estate could be very stressful? Well, I educate my buyers and sellers on how to ease the real estate process making it less stressful and more fun."*

Note: These would qualify as "Elevator Pitch" examples.

To **summarize** my three examples into one final version:

- *"I am an **incredibly useful** and **fun Realtor** to seek when in **need to buy and sell homes**."*

I will then Evaluate the final version to compare against Customer and Challenge in "The 5 C's to UVP." See chapter 3: DISCOVER YOUR CORE VALUES.

1. *Incredibly useful*: Addressing customer's challenge using the "usefulness" value that shows that I am an expert and have advantageous utility to offer.
2. *Fun Realtor*: Invites others to work with me using "fun" as a "uniqueness" trait. People like to work with those who are fun and pleasing.
3. *Need to buy and sell homes*: Targeting home buyers and sellers as those are my customers.

Now, let's **finalize** my real estate tagline. If I am satisfied with my evaluation, I will keep it the same. Otherwise, if I want to target a broader audience—not just buyers or sellers—I may take a step further and come up with this final version:

- *"Real estate is stressful. I make it pleasant and profitable!"*

With this tagline, I made it clear that my industry and target market is *real estate*. I also pointed out the challenge that most people experience when they buy or sell their home, especially for first-timers; the *stress* associated with each real estate process. With *pleasant* and *profitable,* I immediately delivered two reliable solutions to solve people's fears and their financial concerns.

As an agent, you don't have to agree with my final tagline or my perspective on this. My motto reflects who I am and what I believe. It mirrors my values, passion, and expertise. I will use it to connect with others. I will tell my potential buyers and sellers the stories of what made my tagline and who I am today. What you need to understand is that you must have a one-line statement that connects you with others. Create a slogan that reflects your skills, talents, purpose, and passions in life. Come up with a motto and make it your own. Tell people what it means to you, the stories that make you who you are today, and what you want to become. Remember, it is not your tagline that connects you with people; it is "You," and what you can do for them!

Moreover, since your tagline is a mirror of you, and what you do, please don't rush into creating it. You can't rush something that you want to last. Take your time and make it into something beautiful and meaningful. Be innovative and create a tagline that is somewhat intriguing to evoke interest and curiosity. Have fun with it, and make it truly memorable.

"If you want something to last forever, you treat it differently. You shield it and protect it. You never abuse it. You don't expose it to the elements. You don't make it common or ordinary. If it ever becomes tarnished, you lovingly polish it until it gleams like new. It becomes special because you have made it so, and it grows more beautiful and precious as time goes by."

—F. Burton Howard

> *Take the Next Step*
> *Execute Tactic 4.0: Craft Your Tagline*

Craft Your Elevator Pitch

An *elevator pitch* is also known as an *elevator speech*. It's an extension of your tagline that expands on the *proficiency* factor of *"The 3 P's to Elevate the Power of You."* It is usually created to answer the question, *"What do you do for a living?"*

An elevator pitch is often confused with the Unique Value Proposition (UVP). It is similar to a value proposition but without the "why" and the "how" elements. Be my guest if you want to combine your elevator pitch with your UVP. While some coaches do not differentiate the two, I don't suggest you merge them. Why would you voluntarily answer the "why" and "how" questions when you're only asked what you do for a living? You wouldn't deepen the conversation with a new acquaintance you just met out of the norm. A person may be shocked with how well and unexpectedly you've answered, and they often tend to take a step back and build a bearing wall. This act could either be because they felt intimidated, they were just curious, or you left nothing else on the table for them to ask. Therefore, there is a fine line that separates an elevator speech from a value proposition.

When using an elevator pitch, don't go into details about what makes you different and why should others work with you. While you may disagree and pitch your uniqueness, I have had far more success using an elevator speech as a leading step to the value proposition. After someone seems interested in what I do, then, and only then I explain how and why I do what I do differently. With this approach, the introduction feels natural. Also, the conversation is healthy and sounds less like a sales pitch.

Here are some elevator pitch examples:

- *Pablo the Person: What do you do for a living?*
- *Rose the Realtor: I am an **incredibly useful** and fun real estate professional people seek when they need to sell their house.*
- *Ross the Realtor: Do you know how people sometimes need to buy new homes? I am the Realtor that brings them the **most benefits** while keeping the process easy and fun.*
- *James the Realtor: As you may know, real estate can be very stressful. I uniquely help people, so they are **stress-free** when it comes to buying and selling real estate.*

The leading questions would be something similar to these:

- *Pablo to Rose: How are you incredibly useful?*
- *Pablo to Ross: How do you bring them the most benefits?*
- *Pablo to James: How do you make it stress-free?*

By following the conversational concept, you would be having a healthy dialogue. In the next step, which is tapping into your unique values, you would answer how you are unique in what you do. If you were to jump right into the UVP, the other person often ends up saying, "Oh, OK!" "Cool!" or "Wow!" They would have nothing else to ask. Though it is always lovely to make an effort and ask about the other person, the goal here is to keep people engaged and interested in what you do. Otherwise, you've already lost your opportunity.

Furthermore, don't use your elevator speech to pitch your real estate services to others. Use it to engage in conversations and strategically connect with people by allowing them to get to know who you are and what you do.

> *Take the Next Step*
> *Execute Tactic 4.1: Craft Your Elevator Pitch*

Craft Your Unique Value Proposition

Unique Value Propositions are unlike elevator speeches. They serve a formal purpose in business plans, presentations, marketing campaigns, website sales copy, offline promotional material, and other advertising channels. Using the five value characteristics you learned in the previous chapter, create a unique value proposition that will uniquely position you and your business.

Using Rose's example of the first elevator pitch:

- *Rose the Realtor: I am an incredibly useful and fun real estate professional people seek when they need to sell their home.*

When Pablo asks:

- *Pablo to Rose: How are you incredibly useful?*

Rose may answer with the following UVP:

- *"I educate my home sellers through written guides and online videos. This way, they can make educated decisions, have a smooth sale, and a delightful real estate experience. I also have a very unique, top-notch marketing package that gives my sellers a huge advantage in closing their transaction in as little as 30 days. Fast closing translates into more time with their family and more money in their bank account."*

Here are some questions that will help you authenticate the validity and effectiveness of your UVP:

1. Does your UVP reflect who you are?
2. Does your UVP include the values and benefits people receive from working with you?
3. Does your UVP tell people how unique you are?
4. Does your UVP tell people why they should choose to work with you rather than other real estate professionals?
5. Is your UVP less than five statements that customers can fully understand in less than seven seconds?
6. Does your UVP inspire clients to want to do business with you?
7. Have you thought about all of these questions from the customer's perspective?
8. Do you consider every person asking "what do you do" a business opportunity or a potential referral source?

If you feel like you wrote a wordy unique value proposition, do the following:

1. Go back and rethink your message!

2. Remove noise, clutter, and irrelevant information. Leave information that only involves your customers' needs and wants.

3. Rewrite explicit messages that answer customers' problems within seven seconds.

"Everyone's life is a powerful story that can teach us something valuable, but can only be remembered when told with passion."

—James Tyler

Take the Next Step

Execute Tactic 4.2: Craft Your Unique Value Proposition

Your Storyline

You may be wondering why do I need to learn how to create appealing storylines? What does storytelling have to do with my real estate business? Everything! Whether you're telling a story about yourself or a past customer, your services may not appeal to your audience if you can't convey amusing anecdotes.

In real estate, you're dealing with people, and just like any other trade, they want to trust you before they do business with you. Though potential clients wish to learn about what you can do for them, they are more interested in you. They are interested in your background, where you came from, and how you got into real estate. Did you grow up in town? How did you help others in the neighborhood, and what was so distinct about past circumstances? They want to hear your backstory, recent events from which they can learn, or new moments to which they can relate. Share backstories of similar situations of your audience so they can connect, and get the urge to work with you.

While some agents miss this—or not give it much of substance, a little storytelling and a bit of background information go a long way in building relationships, trust, and credibility. People love stories—as they give a sense of comfort. Stories also make you memorable and add the uniqueness element in your business. Telling charming and engaging anecdotes will surely differentiate you from most real estate agents competing in your area.

Storyline Approaches

Use the following six approaches to help you write an original storyline:

1. Loss and Redemption
2. Before and After
3. Us vs. Them
4. Valuable Discovery
5. Third-Person Testimonials
6. Secret Telling

Loss and Redemption

Starting your storyline with a loss and redemption situation is very powerful in attracting people's attention. Disaster and recovery adventures can quickly stir people's emotions as they show the upside of going through hardships and encountering challenges. Don't be discouraged from sharing a loss and redemption experience in your story if you've had one. For example:

I had everything I wanted; I had it all. I was on top of the world. Then [some loss] occurred that changed my life. I had to figure my way out of [the consequence of failure], but it turned out to be a blessing in disguise because I went through [some life lesson or positive experience], and I [learned or received some value]. Now, I [recovered, positive outcome, or am grateful to be a Realtor].

Before and After

The comparison stories of transformation are the most commonly used stories by influencers and top real estate producers, and they work great in any market. How many times you have heard millionaires say *"I came from nothing," "I dropped out of college," "My family was poor," "If I can do it, you can too."* The *Before and After* tactic is supposed to motivate listeners, and show a value in the transition thereof. In this approach, the heavy focus is on the transformation, which could be hard work, dedication, consistency, or whatever bizarre, unconventional actions led to the desired outcome. For instance:

"I was poor living with [family or friends]. Then [the transition or transformation, did something of value, worked hard for years, or learned real estate]. Now I have a [desired outcome, benefit to the customer, or became the top real estate office in Hollywood, ranked #1 in customer service by the Los Angeles Times]."

Us vs. Them

Reveal confidence in your opinions so you can build loyal customers, followers, or what Seth Godin calls, your "Tribes." Don't be the relatable person who is bland and neutral on many topics. Don't only share safe facts with which everyone agrees. Harness the power of polarity and share your opinions on hard matters, and stick to them no matter how many people disagree with you. Create your world, be real, and tell it how it is; let people fit into your life instead of trying to fit everyone into yours.

Use *Us vs. Them* stories to polarize your target audience. Using this type of approach will draw your audience closer. Whether they are buyers, sellers, partners, or friends, you want these people to choose who they are to you. If they're with you, then they will be the ones who have your back no matter what. Those will continue to ascend with you regardless of the real estate market situation or what happens in your life. Here's an example:

Top real estate agents like me, like to work with sellers who appreciate "us." They tend to understand our dedication and hard work, so they allow "us" the opportunity to serve them. Not only do they love working with "us," but they also refer "us" clients and keep using "us" in the future because they know we have their best interest in mind. Average agents reduce their commission to get a listing. We don't because our benefits exceed theirs. Therefore, you will experience a vast difference working with "us vs. them." Are you willing to take such a risk of working with "us vs. them?"

Valuable Discovery

Going through life, we are continually learning new lessons and discovering new values. We often need to stop, take note of those values, and share with others—especially our tribe. That is why leaders scale and become even more powerful. They are regularly providing benefits they discover. Their storyline keeps on evolving and ascending. How many times have you read stories about Steve Jobs or Bill Gates? Here's a perfect example of a valuable discovery.

"Oh my gosh! Wait until you hear about this new thing I just discovered! You're not going to believe it, I thought it was a rumor, but it's true. Did you know that we can now have a robot at every open house 24/7 answering potential buyer's questions?"

Third-Person Testimonials

In your storyline, or while having conversations with clients, drop the names of people you know or past clients who said credible words about you or your services. It shocks me that most real estate agents, especially new licensees, think that testimonials are only valid if they were from past customers. Testifying, whether you are a person of integrity doesn't have to come from customers only. A family member could give you a "Yelp" review about how dedicated you are. A friend may talk about how many years you've been in the industry. A mentor can speak about your leadership traits and your high level of commitment. Your spouse or relatives can converse about how caring you are.

You can genuinely get a dozen of real people to speak of you without having a single customer. You are already a success; allow people to share that social proof about you. Highlight that power in your storyline. That is what customers want to see, read, and hear.

"Did you happen to read some of my Zillow reviews, especially the one written by Pablo Petersen? You might also like the video testimonials Frank Ferrell, and Sally Smith recorded on YouTube about how delighted they were with my services."

Secret Telling

Have you had a giggle telling your spouse about a little secret you had going on while you were in high school? Maybe about you and your friends and how you were the guy behind the "Chair Scare" using an air horn?

Telling our secrets sometimes makes it easier to see where we have been in our lives and where we are going. Telling our secrets also opens the door for other people to tell us a secret or two of their own. Exchanging stories with little secrets is what being a human is all about. It is also an act of trust and confidence. Allowing others to enter that deep layer within us, where we keep our secrets, brings them closer than we do anywhere else. Tell secrets in the form of stories to connect with people; however, don't tell secrets about other people.

I've got a secret! If you want to find out what it is; you need to finish this entire book. Here's a hint, it is in one of the last chapters of the book. Find it, and I will guarantee you will dominate.

Storyline Characteristics

Now that you have the six storytelling approaches to write your storyline, how do you elevate the power of your storytelling tactics? Allow me to introduce these five characteristics that will keep your audience on the edge of their seats, charm them, and leave them wanting more: Character, Circumstance, Conflict, Curiosity, and Conversations.

Character

A *character* is an actor or persona that attracts clients or customers. This identity could be the sympathetic "You"—if you bought or sold before—or someone that your customers can relate to—e.g., another buyer or seller.

A character in a story should be a reflection of your audience. It is someone that had the same circumstance as your current customer. For example, a past client whom you've helped fulfill his or her needs. When you tell a backstory using the character as the example, you leave your listeners with the mindset that they want to be like that past client. Using personas to create clear visions of the success journey potential customers would experience, help ease their decision of choosing to hire you.

For example, suppose you're a selling agent representing a buyer with a contingent contract to sell within a specific date. The buyer may lose the earnest deposit if the sale doesn't close on or before the contingency date. You can tell a story about the previous *character*—that is the other buyer you represented in a similar circumstance—and how you handled the situation. This way, your current buyer can relate, learn, and trust you'll deliver.

I am pleased to introduce to you Bob the Buyer, Barbara the Buyer, Sally the Seller, Steve the Seller, Frank the FSBO, Ellie the Expired, Rose the Realtor, Ross the Realtor, James the Realtor, Ron the Renter, Fred the Friend, George the Grantor, and Pablo the Person. I added these *characters* through stories and examples to which you may relate, and also to turn this somewhat tedious writing into an exciting journey. These are the people that will accompany you along the "Dominate Real Estate" expedition.

Circumstance

Relate to your audience by talking about past customer experiences. Teach people core principles that you want them to understand whether being about life in general or the real estate business. Most people let life pass them by, and they don't slow down to take note of interesting things that happen to them. You are different. You can use circumstances that occur throughout your life to teach and inspire others—as well as sell real estate.

Don't just brag about how many homes you've sold, or your continued success, or how you're the number one agent in the area. Start talking about failures and how you went about overcoming them. We all know life isn't perfect, and everyone knows how stressful the real estate process could be. Talking about flaws is what real estate agents struggle with sharing. However, it is one of the most crucial tips in sharing stories because it makes you relatable and real. So, don't try to sound like you're the perfect real estate agent. When telling a story, people want to hear about the chaos, your flaws, and your failures. Why? As human beings, we relate to your failures because we are all flawed. People will connect when you talk about the mess and end your story with success.

Conflict

It is acceptable to have a dispute and disagree when reasonable. People don't always know the right thing to do or the proper step to take. They seek your real estate services so you can help them decide, walk them through the process, and ease their pain. None of your real estate facts and figures matter until you have established some emotional connection with your audience.

When crafting a story, spark the emotional side of your audience's brain. Think about what emotions you want to communicate and then provide information to support those emotions. Stories are a great way to connect with your customers emotionally. Help your customers make better decisions by telling them stories that are relevant to their emotional conflicts.

Customers will connect with you on a much deeper level when you polarize your opinions. You want them to be on your side. Delve your customers into your stories, engage them, and get them to the level where they start using "us" in their sentences. Don't be a tedious agent with a neutral opinion at all times trying to please and win everyone. When asked, give your honest advice that is in your customers' best interest. Most will agree, and some won't. You can't win them all. Create die-hard customers who will believe and trust in you, follow your advice, share your message, and do business with you over and over again.

Curiosity

Stories are exciting, and much more impressive and enjoyable when you start your backstory in the middle. Far too often, storytellers, salespeople, and marketers give way too much detail upfront. They begin their story in chronological order, putting the audience to sleep before the exciting stuff occurs. By the time they've reached the "Ah-ha" moment, their audience has already lost interest.

A good curiosity tactic you can use is to divide your story into a few mini-stories. Before you transition between mini-stories, ask a question but don't answer it right away. Pause, start another mini-story to answer the question you just asked. Before ending the current mini-story, ask another question, then go into another mini-story to answer the second question. Throw a sense of curiosity in your stories to keep your potential customers on their toes until the end. This curiosity tactic is called "Story Loop" in storytelling. It is one of the most challenging tactics to master, as you must keep track of which loops are open, which you're closing, and which ones are about to open.

Conversations

Get your customers engaged through attractive, easy to remember little stories to illustrate a relevant point. For instance, captivate their attention through senses to inspire them with reality. For example:

"Don't worry Sally. Bob, the home buyer, will smell the freshly baked chocolate chip cookies rushing into his nose the second I open the door of our open house. He will enjoy the aroma of Napa's red wine in the backyard while tuning into the relaxing piano sound of George Winston playing in the background. Rest assured our open house experience will be different and beyond expectations. Wouldn't you agree?"

Tantalizing the senses through your story engages your client immediately. Set the scene by describing how it would visually appeal. Describe the sounds that will occur and the pleasant aroma that will fill the air. How did the scenery make others feel? Engross your audience into your enchanting stories through the experience of the senses.

Get your audience emotionally invested by following the principles of VAKO: Visual, Auditory, Kinesthetic, and Olfactory. By using these four elements while telling your stories, you're likely to lure in and lock down your audience. Make sure to give your audience what matters, keep them interested, but don't overdo it.

To conclude, stories are a great way to communicate and connect with your audience effectively. Put everything you've learned so far into an inspiring backstory. Why real estate, what's your purpose, and what's your passion? Who are you, and how do you want people to view you? Whom do you serve, and what do you do for them? What makes you different in what you do? How do you benefit your community?

How do you make people feel better about their decisions? Use your creative imagination, your tagline, elevator pitch, and your unique value proposition.

Emphasize on your personality to craft a storyline that will inspire people and entice them to want to hire you as their real estate agent of choice. Make your story about people, animals, real estate, or whatever your passion is. Touch people's emotions and make sure there is merit to the storyline. Your story reflects who you are. Make it count.

> *Take the Next Step*
> *Execute Tactic 4.3: Craft Your Storyline*

Your Biography

Write your biography for everyone around the world that wants to know a little background about who you are, your education, certificates, awards, interests, and your hobbies.

The purpose of your biography isn't to sell, promote, or show value in the services you provide. People who do not know you will be reading your bio. Since those people are newcomers, you want to have a mean that invites them to learn about you. Your memoir does that; it encourages others to connect with you and welcomes them into your world.

If you want to dominate your real estate market, you are very likely to work with people who have never heard of you or met you. Your biography is a useful tool that will connect strangers with you on a personal level and eventually you with them—then soon to be customers.

Share some personal information about your greatest passions, your natural talents, a little about your spouse, kids, or pets. Also, the more you know about in the different fields of sports, interests, and hobbies. You can also write about your favorite movies and TV shows or Podcasts. Perhaps you can write about the unique talents you possess since childhood, compliments you often receive, and the top three unique things people remember about you. Attracting people with common similarities and interests will provide them a reason to approach you and work with you.

You want to give readers something for them to talk to you about if they meet with you. Tell them things to connect with you through, and something by which to remember you. Do you want to be successful in real estate? Start building relationships by introducing yourself to your community. Allow them the opportunity to look you up, find you, and get to know you, early on in the process.

Keep your biography simple and clear. Avoid clutter or noise. Most importantly, let people know the best ways they can get a hold of you.

> **Take the Next Step**
>
> *Execute Tactic 4.4: Craft Your Biography*

Your Executive Summary

Write a professional executive summary that targets partners, vendors, professionals, and referral sources. Elevate your business and your professional values to your peers. Begin your executive summary with your unique value proposition. Talk about your vision for what you hope to achieve in business and life. Talk about what you did, what you do, whom you serve, and why. What are your business goals, mission, and vision? What are your objectives and keys to success?

Your executive summary must show your talents, the different roles, and positions you held in all businesses. Address your Sphere of Influence (SOI) and partners by profession within your summary. Ask for business and thank them in advance for referring potential clients to you.

Your executive summary is a short version of your business plan. Anyone reading it should grasp the kind of business you operate. It should explain the sort of service you offer, clearly identify your target audience, the benefits of your products, and the type of referrals you wish to receive. Don't forget to provide your contact information and tell your audience how you want them to engage with you.

> **Take the Next Step**
>
> *Execute Tactic 4.5: Craft Your Executive Summary*
>
> *Execute Tactic 4.6: Rank on the First Page of Google's Search Results*

Congratulations! You are now on your way to be part of the top 3% of agents in your town.

CHAPTER FIVE

SPEAK YOUR MESSAGE WITH INFLUENTIAL CONTENT

"Earn massive credibility and strengthen your brand by educating the people you wish to serve."
—James Tyler

A mateurs spend money on building their so-called remarkable websites. They spend more money on paid ads to get people to their sites, but no one calls! They might think the quality of the traffic was inadequate. They could believe that the design of their website was the issue, or people weren't interested, and the leads were terrible—when in most cases, the problem is content.

The first step in developing a relationship with a potential customer is through a clear line of communication. Most sales copy and websites are usually confusing. The message they try to convey isn't transparent enough to get people to take action, reach out, and become clients. The same is true for email templates, social media posts, and SMS messages.

Furthermore, most businesses offer more than just one product or service to their clients. Quite often, those businesses try to answer ten problems a client might have in one solution. The abundance of product features overwhelms people and gears their focus away. Visitors eventually end up leaving the brand without buying due to lack of clarity in services offered.

Now combine sparse and confusing material with an overwhelming features-jammed solution; the result is a high risk of losing every potential customer comes across such a brand.

Compose Clear Brand Messages

The top-ranked real estate professionals are known for their credibility, clear communication, reliable personal brand, and excellent values. In this chapter, you will learn how to write clear and concise brand messages that will entice people to reach out to you, and hire you as their real estate agent of choice.

Communicate Value First

In the third chapter of this book, you learned "The Five Value Elements" that elevate the perceived value of any entity, including you as a brand, and your real estate services—Demand, Utility, Scarcity, Flexibility, and Credibility. You also learned about "The 5 C's to UVP" that will unquestionably help you craft a compelling, unique value proposition for any product including your real estate services—Customer, Challenge, Competition, Compound, and Close. The goal was to communicate your benefits to your target audience intuitively, and the people you come across. Hence, adequately answering the questions, *"What do you do for a living?"* and *"What makes you different?"* Well, the same is true for people and technology.

Since a vast majority of today's communication happen online, technologies such as your websites, social media profiles, and email newsletters must have your unique value propositions.

Microsoft conducted a widely publicized study concluded that the average American attention span is only 8 seconds—less than a goldfish. Whether online or via phone conversations, if people have to think past eight seconds to understand how your brand solves their needs; you've already lost. Your message must directly and effortlessly touch on their wants within the first eight seconds.

To write a satisfactory sales copy and not confuse potential clients, don't try to answer ten problems in one solution. Build a unique value proposition for every product you offer, showing each product's benefits to each target audience. Apply your unique value proposition to articles, blog posts, social media posts, paid ads, and emails newsletters.

Whether in person, offline, or online—the more you elevate the values of your content, the more likely you will influence, convert, and monetize your potential audience.

Clarity is Key

People usually don't realize that their messages are not 100% clear. If your communication isn't clear and isn't directly helping your customers solve their needs; they aren't going to pay attention.

Customers will only listen once you clarify your brand message. They pay attention when you expose them to a brand that unquestionably touches on their pain points, and explains the solution to their problem. People typically don't just buy the best products or services. They purchase products that they understand, why they need it, and how it benefits them.

In most cases, businesses fail because they invest in the wrong domain. They spend so much money on promoting the company when they should be investing in their message. For instance, "Apple" does a great job at writing sharp brand messages to show the benefits of its core products. That is because people buy solutions, not ads. Write brand messages that portrait people's situations and their lifestyle after your brand solves their problems.

Clarity is key! Skip through the fluff and the noise. Offer your values in simple ways to be processed by the human brain within seconds. Win your marketplace with crystal brand messages. Write enthralling sales copy with clarity in your purpose and your business. Make clear why customers need your real estate services. Your success is a race of your potential customers understanding your products and the benefits you offer. You confuse them; you lose them!

Credibility is a Must

Stand out from your competitors, influence your target audience, and close more sales by building trust and credibility.

Build Credibility Through Content

Establish yourself as a well-known and friendly expert in the real estate market. It is with no doubt that you can positively influence your target audience with valuable information. Gain trust and authorship by educating your customers.

By developing fascinating, relevant, and useful content, not only you will connect with more people, but you will position yourself as an expert in your local community. The more people that will consume your advice, the more relationships that you will build, and the higher your position is in the real estate industry. Thus, you will earn more trust and credibility.

Use reliable communication to stay in touch, nurture relationships, and deliver on your customers' needs. Your keen real estate knowledge will continue to position you as the expert in your space, and therefore influence people's decision-making in choosing to work with you.

Build Credibility Through the Sales Cycle

Build meaningful relationships and earn people's trust over time through the real estate sales process. Review your sales process and design a sequence of phases or milestones that split the process into multiple stages. Position your brand messages to build trust and communicate effectively with each one of your potential customers throughout each step.

Understand your customer's needs in each step of the sales cycle and address them at the right time with the right message, you will smoothly and conveniently move them through the real estate transaction and gradually earn their trust every step of the way.

Build trust by having a clear view of the sales process, its stages, and needed actions. Send personalized messages, follow-ups, follow-through, and move people a step forward each time until they become loyal customers.

Context is Powerful

Now that you understand the importance of value, clarity, and credibility, the one thing that rules them all is context. Your brand message must be meaningful unique in style. Chosen words are the first key elements that shape people's decision on whether they will continue listening, reading, or watching your message. Choose your words wisely.

Brand personality and looks matter! Setting the tone, colors, voice, and the image of your brand is also very important in the effectiveness of your message. People respond to different styles of communication. They are more likely to respond better to authoritative or educational information than a sales copy. Others may prefer conversational over inspirational content. Don't limit yourself. Spice up your content with different styles and multiple formats to test what material performs best with each specific target audience.

In conclusion, write a clear message that reflects your brand, so it is highly effective, and it answers to your client's needs. Practice your craft and genuinely come up with solutions. Be spontaneous, optimistic, and funny. Be amazing at what you are selling and speak with complete confidence. Build trust, deepen your voice, straighten your posture, and looks. Be animated and let your ambition go. Whether in person or online, allow yourself to express yourself. Remember, you are the brand.

Craft Persuasive Headlines

How do you get customers' attention, and how do you get them to work with you? When writing your brand messages for online consumption and utilization, headlines are the first and foremost important factor to consider in getting your potential customer's attention. Headlines are the first opportunity and the most significant chance you have to invite your audience to learn about your brand and services. They determine whether or not your target audience is going to be attentive and read the rest of your brand message.

Grab Your Audience's Attention with Headlines

Studies show that 80% of the decision-making on whether someone would read an article or watch a video is made primarily based on the pertaining headlines. To turn people into potential customers, you've got to turn them into potential readers first. Thus, it is highly critical to write compelling headlines that will grab the attention of the audience you are trying to reach, primarily online.

With the amount of content reproduced on the web, the average online surfer will scan through about 80% of headlines, but only about 20% of those people will read your message. Thus, it behooves you to invest just as much time perfecting a solid headline that gets people's attention as you do on the content within your message.

The way you use headlines can pull or repel readers. Think of them like magnets; if used correctly, they can draw your audience to read your brand message. Otherwise, they will push them away. A headline is the first line of content your target market sees and understands when they come across your brand. If it's interesting enough, they will likely to click and read or watch. Otherwise, they'll go away!

There is an art in writing catchy and irresistible headlines. Compelling headlines will position your articles in a way that will enthrall your audience's attention. Titles should identify the context of your document, its relevance, and how valuable it is to your audience. They will affect your message readability, searchability, and virality. Also, captions will affect your brand performance and lead acquisition. To get people to listen and buy from you, you've got to get their attention. To get their attention, you've got to write captivating and convincing headlines.

Using headlines as the magnet that attracts your target market's attention could genuinely be the stepping stone to your business's success. They can significantly increase the chance of virality. Writing headlines that will entice people into consuming your content, and sharing it with thousands of people will build credibility and position your brand among the best. How do you write headlines that do not push your audience away, but instead pull them to rave about you, and your brand?

Create Powerful Headlines

Every great marketer and salesperson knows that grabbing people's attention and increasing their interest remains paramount in the business world. It is imperative to spark curiosity and drive people to learn about a particular brand, products, or services. Studies show that the overall success of all marketing and sales campaigns rely heavily on the quality of their headlines!

Think about it. You craft compelling headlines when you write blog articles to get people to read them or post on social media to get people to follow you. Producing a YouTube video requires several iterations of the headline to entice others into watching it. Further, writing a white paper to get potential leads or an engaging eBook title to educate your customers, and composing a subject of an email to get subscribers to open it. As you can see, with every aspect of your online marketing channel, you must think of captivating headlines that will get your clients attention.

Here are the top three components that will increase readership and help you create compelling headlines:

1. **Customer:** Get to know your target audience and what makes them tick.
2. **Challenges**: Seek to understand their needs, wants, and desires.
3. **Community**: Look into what they share online, whom they follow on social media, and in which groups they participate.

Knowing your target audience, their needs, and what interests them will help you draft headlines that speak directly to their challenges. We have covered much ground on this.

Generally speaking, most people are interested in articles or posts that provide a useful list of things. Making "lists" tend to simplify complex information that people need. By creating a list of easy steps to follow, you are more likely to drive people to read your letter.

Similarly, "How-to" information teaches people something that helps them achieve the desired outcome. "How-to" is one of the most searched keywords on search engines such as Google and YouTube. By producing "how-to" articles and videos to answer your audience concerns, you gain a significant share of your marketplace.

Furthermore, besides using headlines that increase usefulness and quality of life, you get people to read your headlines by adding attention-grabbing words that have the power to touch emotions and inspire desire, fear, anxiety, curiosity, joy, happiness, surprise, and shock.

Keep up with the latest industry news and trending topics that solve your customer needs, boost their productivity, or help people achieve higher success and more profits. Tell a little story and add an infectious agent to your headline and watch your message go viral.

Writing compelling headlines isn't rocket science—once you understand what causes people to click, read, and share. The critical point to remember is that people consume content because of what it shows about them and their desires. Again, start with the needs of your target audience, and then layer in the concepts revealed above to write gripping and captivating headlines.

As with everything else, writing compelling headlines gets simpler with practice. Although it may seem unnecessary and tedious at first, compelling headlines are the differentiating factor between reaching your potential customers and losing your intelligence in the chaos of the internet.

> *Take the Next Step*
> *Execute Tactic 5.0: Create an Educational Branded Guide*

Here are a couple of examples of compelling headlines you can use today:

- Wouldn't You Be Thrilled to Find out How to Buy 10 Homes in Less Than 12 Months?
- 100 Reasons That Will Stop You from Renting and Enjoy Owning a Home Next Month
- 30 Tips That Will Inspire You to Buy and Sell Real Estate like a Pro
- 20 Secrets and Techniques All Sellers Should Know Before Selling Their Homes
- You will Be Delight by the Exterior, but Captivated by the Gourmet Kitchen!
- Soak in the Daily Joy of the Sunset with This Stunning

Oceanfront 4-Bedroom Home!

Create Compelling Calls to Action (CTAs)

Calls to action such as links, buttons, or invitations that pursue others to take action are the sole indicators whether people will proceed to engage. Calls to action are essential attributes that will lure people in, woo them into taking the next step, and eventually, working with you.

Persuade Your Audience Into Action

A call to action or CTA is a foundational marketing element that is designed to persuade someone to take action. Calls to action are used to get readers or visitors to do something after connecting with your brand. For example, sign up for an email list, download a document, a follow request on social media, share an article, watch a video, or sign an agreement. Whatever your call to action may be, the goal is to entice your audience into engaging with your brand and eventually reaping the rewards by converting them to customers.

Create Clear Calls to Action

Integrate calls to action in both offline and online channels; also, through each department within your real estate business. When doing so, you would think it is evident that your audience should know what you want them to do next. However, in reality, it is not always that clear neither that easy for them to figure it out. Although you expect potential customers to know why they're interacting with your brand, you have to spell out what you want them to do next clearly. For example, adding a phone number to a For-Sale sign and assuming potential buyers will call when you want them to send a text message instead. Be specific and transparent with your request: "TXT the number for more info."

The next step they need to take is usually unclear and mostly vague to prospects. Web designers and marketers spend an immense amount of time thinking about the various elements of a website or a piece of assets such as images, layouts, colors, and form fields. However, often, they skip over the calls to action, which could create a burdensome for readers to follow through. Taking that next step can sometimes be the difference between a conversion and a loss of a lead.

Your call to action is an essential element that propels your audience through your brand messages. It can act as a gateway to the next step of your marketing or sales process, where every move could lead to a potential customer. Don't wait for sellers to ask what's next! If you finished your listing presentation, ask them to sign the contract.

It's easy to leave your CTA button saying "Submit." However, it should have a concise meaning, and it should always represent an explicit action your visitors are bound to taking next. Here are three elements of persuasion that you can use in your calls to action:

1. Persuade with Benefits
2. Persuade with Scarcity
3. Persuade with Exclusivity

Persuade with Benefits

As you've already discovered, words are mighty—only if they're used to speak directly to your customer's needs. With calls to action, the same is true. Learning how to use the right phrases in your calls to action can dramatically increase your conversions, and therefore, your pipeline. Understanding why specific phrases are more effective than others and learning how to implement them will set you up to become a professional at creating copy that converts and translates into customers.

When creating a call to action, you're trying to tap into the audience emotion and get them to complete a viable response. Get customers to follow through by using words that convince and persuade them within the right context. Engage them by creating the calls to action that focus on the benefits, not your services. Your call to action is all about driving the readers to learn more about what values you bring them, not to tell them about the products you are offering. Focus on the results the reader will get and the action they should take.

For example, a button that reads "Free List Price Guide for Sellers" is much more valuable than "Download" or "Submit." Sellers can see the benefits they get right away.

Try mixing these phrases into your calls to action when using persuasion:

- You
- Guarantee
- Results
- Free
- Because
- New
- Discover
- Find out

- How to
- Learn

Persuade with Scarcity

Trying to convince, motivate, and urge people into doing anything could be a difficult task. Especially if what you are asking is a big step or a financial obligation. Waiting for the right opportunity seems always to be the easy way out. Typically, people don't rush into buying things; they take their time and often wait for a price reduction. More often than not, tomorrow becomes eventually, which ends up being never.

Creating a sense of urgency gets customers to take action immediately. The sooner you can get people to take action, the better. Persuade your target audience by creating a sense of urgency and scarcity. For example, instead of using "Open House," try "Open House. Today Only!," or "Open House. Hurry In," or "Open House. Last Chance for Offers!"

Use these practical calls to action phrases that will make people rush to grab whatever you're offering:

- Only X days left
- Limited supply/inventory
- Closing soon
- While supplies last
- Today ONLY!
- Last chance
- Offer ends on "date!"
- Hurry
- You're Running Out of Time!
- Immediately
- Now

Persuade with Exclusivity

Another compelling way to persuade people into taking action is to imply exclusivity. Limiting whatever you're offering makes people want what they can't have. Besides, exclusivity makes people feel special. There's that element of a community characteristic because people want to be part of a distinctive group. However, how do you make something look and feel more exclusive? Of course, with a prominent call to action!

Try these phrases out to get people eager to be a part of your products or services:

- Request an invitation

- Now closed
- Members Only
- Subscribers only
- Only available to "[seller/buyer]."
- Pre-register
- Pre-order
- Limited spots
- Exclusive access
- Reserve your spot now!

Create Gauging Calls to Action

Sometimes the action you want your audience to take isn't necessarily to sign-up for an email list or use a service. Perhaps your goal is to measure the interest of your audience in a particular effort. For example, before boosting an ad on Facebook, you could create the social media post and test its effectiveness organically. If your fans seem to like that post and engage with it; then it's a good indication that it might be a good fit for a paid ad campaign. This technique can help you build even more anticipation around similar strategies.

While gauging calls to action could be useful in some cases, they are not very effective in persuading others into taking action. Use a benefit, scarcity or exclusivity call to action type in addition to your gauging call to action.

Here are samples of excellent calls to action that will help you see how interested and engaged your audience could be:

- Reply
- Like
- Comment
- Share
- Follow
- Contact me
- Apply Now
- Start the quiz

Transparency and open communication in business have always been effective key selling elements. Calls to action are no different. Be transparent and use them as a communication tool—just like headlines. The more open you are to listening and engaging, the better your calls to action and the more your audience will appreciate interacting with your real estate brand.

It's time to go beyond merely having a contact form buried at the bottom of your website. Embed your calls to action within your blog articles, comments, emails, and social media posts. Don't just be where you want people to reach out to you, tell them how to reach you, and what actions to take. Doing so will show you how active your following is, and build a strong community around your brand.

With every brand message you distribute to the world, don't be afraid to reach out to your audience and invite them to contact you. You might be pleasantly surprised by the responses you get.

There are thousands of call-to-action phrases that you can implement to drive more engagement and interest to your brand. Start experimenting to find out what works best with your audience.

> *Take the Next Step*
> *Execute Tactic 5.1: Create Branded Guides*

Here are some example headlines that you can use, edit, or improve to help you get started. Rewrite them in different styles to target multiple markets. For instance, for sellers, buyers, and first-time home buyers.

- Everything You Need to Know About Real Estate for [Sellers/Buyers/FTHB]
- From Start-to-Finish Real Estate Guide For [Sellers/Buyers]
- The Complete Journey With Step-by-Step to [Selling/Buying] Your Home
- A Complete Step-by-Step Guide on How to [Buy/Sell] a Home in 30 Days!
- How to Quickly Get the Highest Offer on Your Home for Sale
- The Best Methods Buyers Can Use to Find Perfectly Priced Homes for Sale
- How to Make Your First Home More Valuable and Sell Faster
- The Secrets to Sell Your House for the Most Money
- The Comprehensive Guide to Price Your Home to Sell

Content Creation Steps

Remember also to apply everything you've learned in the previous chapter about your target markets and your unique value proposition when building your content. Here are some steps you can take to build highly marketable letters that will make your target audience rave and babble about you:

1. Create a guide for each specific target audience.

2. Craft, edit and finalize a highly effective and compelling headline.
3. Determine one main challenge a target audience is looking to solve.
4. Write about one solution that you can deliver to your target market, and solve their challenges.
5. List all of the benefits your target market will experience as a result of your content.
6. Include your UVP to tell your audience why they should work with you.
7. Include links to your Bio or Storyline and your Executive Summary. Position yourself as the real estate expert in your marketplace.
8. Add a persuasive call to action. How do you want your target audience to interact and engage with you?
9. Save the guides into enticing PDF documents to offer on your website.
10. Ensure that you have an opt-in form displayed prominently on every page of your website to download your offers.

Note

There is more to discuss on the last two bullet points in the upcoming chapters.

Establishing a line of communication is the first step in developing a relationship of trust. Influence your potential customers and start building credible relationships through your branded guides.

Content Creation Process

1. Content Research: Start with Google search. Study the competition, find top keywords in your industry, validate demand, look on FB Groups, Quora answers, and look for information gaps that were missed by your competitors.
2. Content Idea: Write about customer's problems, relevant industry news, and trending topics
3. Content Outline: Create a Table of Contents to keep your thoughts and notes organized.
4. Add Viral Elements: Mention top influencers or include relevant quotes from leading influencers. Include reference data, stats,

research paper, and survey to support your facts.

5. Add Promotional Features: Install the Click to Tweet feature and Share Icons.

6. Create Visual Media: Take photos, design illustrations, or produce videos.

7. Content Writing: Write 800 to 2000 words articles.

8. Add Mixed Media: Embed videos, Instagram images, slide decks or slideshares, infographics, micrographics, and gifs.

9. Content Review: Proofread and finalize your papers for publishing.

Content Creation Tips

Use Quuu and BuzzSumo to find trending content by keyword or topic. Find new information that already has social proof. Quuu will discover information with moving subjects. BuzzSumo will show you articles and stories that have the most shares on social media. Combine the two platforms to write compelling messages, and trending content people have the propensity to share. You can then repurpose them for your particular needs and advantages. Rewrite trending articles with your brand and add the calls to action to capture your audience's attention and their contact information. For example, this report "Zillow CEO Spencer Rascoff sold his home for much less than Zestimate" could make a notable article on your blog. You can share it with your potential sellers who trust in Zillow's Zestimate to determine the price of their home. Thus, elevating your position as a real estate agent who could provide a more accurate value than Zillow!

Take the Next Step

Execute Tactic 5.2: Curate Trending Content Using Quuu and BuzzSumo

Create Quality Brand Content

While it is critical to write explicit messages that catch your audience's attention, it is also vital to create messages that resonate with their needs. Your content needs to speak to your clients through each stage of their journey interacting with your brand.

How will you write your brand content, what methods you use to deliver the information, and how will your potential customers consume it is imperative. Choosing the proper type of material and the right communication channels have a critical role in moving leads from cold to hot, and then to customers.

There are three stages that you need to be aware of within your customer's journey—the awareness stage, which falls at the top of the marketing funnel. The nurturing step, which is in the middle, and finally, the conversion stage falls at the bottom of the funnel. Your content needs to resonate with your audience at each stage of their journey to be effective at converting them from strangers (top) to customers (bottom of the funnel).

Keep in mind these five steps when building your brand content through each stage of the funnel:

1. **Determine** your target audience and the purpose of the information. For example, generate buyer leads or convert sellers to customers.

2. **Discover** the type of messages that you need to produce— for example, website content, downloadable material, email templates, social media posts, or text messages.

3. **Define** the required resources to develop your data. For example, sales copy, headers, calls to action, reviews, and charts.

4. **Decide** on the level of flexibility and scalability of the information you're going to create. For example, will your target markets read, listen, watch, or consume such content in-person? Can you repurpose your assets from, e.g., PDF to video?

5. **Distinguish** between the kinds of solutions you will offer your target audience. Will the content solve a need (functional benefits), a want (physical advantage), or a desire (emotional interest)?

Create Content to Build Awareness and Drive Traffic

You can only show up to open houses and customer appointments so many times a week. Hence, why it's crucial to develop content to drive traffic to your brand and spread your message while you're unavailable.

The biggest traffic driver on the web is a blog. If you don't have a blog get one. Fast! Most real estate companies nowadays support their agents with already-made WordPress websites. If not, hire a professional. Investing in a professional-looking and behaving blog is one of the most important and most rewarding investments into your real estate business.

Every large enterprise and every top real estate producer in the nation has a blog. They all produce data weekly. The more content they generate, the more exposure, traffic, and leads they receive! The longer you wait, the more successful they will become. The quicker you'll build a blog and produce reliable advice, the faster you'll begin capturing some of their audience. Build a blog, so you can create awareness and drive traffic to your brand.

Social media has become the number one channel to build brand awareness. Large enterprises use it to drive high web traffic to their brands. Surprisingly, most real estate individuals and companies don't leverage social media correctly. Here's why: They spend so much time engaging in other people's social media profiles, so much more than investing in their own branded assets. Instead of focusing on guest blogging, or referral traffic from partners and other people's social media profile, as a real estate agent, you must direct your focus on producing your content on your blog, share it with the world, then engage to drive traffic back to your brand.

Build awareness and drive traffic by thinking of your blog like your business hub where your customer's journey begins. Write articles, record podcasts, and produce videos to share on your social media profiles. Then engage on your shared content to bring people back to your blog. Did you know that providing value and genuinely driving traffic to your sites generates the best quality leads?

For example, if you share an article from your blog to Facebook or LinkedIn, and people note their remarks, you engage, and drive them back to your blog. Similarly, if you share a video to Facebook or YouTube, and people comment on it, again, you push them back to your blog. This way, you build awareness and drive strangers from social media profiles to your website. Once they are on your site, you can then capture their information and nurture them to become customers eventually.

Here are the top organic—not paid—social media sources to build awareness and drive traffic to your blog or website sorted by effectiveness:

1. Facebook Groups
2. Facebook Messenger
3. Instagram Direct Messages
4. LinkedIn Groups
5. LinkedIn InMail
6. LinkedIn Pulse
7. Medium by Twitter
8. Google My Business

9. Pinterest Pins
10. Twitter

> **Take the Next Step**
>
> *Execute Tactic 5.3: Create Branded Content to Build Awareness and Drive Traffic*

Create Content to Capture and Nurture Leads

Information products like ebooks, presentations, audio recordings, and videos will strengthen your brand message. Every time people consume them, they reinforce your promise and your expertise, even while you're away.

These tools are called lead magnets. They are used on your websites with lead capture forms to obtain your audience contact information, nurture relationships, and build trust. Marketers call this process "micro-conversions," which transition your prospects from visitors to leads and eventually from opportunities to customers.

Branded valuable information, especially in the form of PDF Guides, is the ultimate way to leverage your time and your marketing efforts. They speak on your behalf without limitation and for many years to come. Many of my real estate customers use this method to grow their business. The BREL Team, a real estate broker, based in Toronto, generated over 80% of their $150M annual closing in 2018 from one single blog built in late 2014. Branded valuable content is king. It works! Use it to capture and nurture your leads immediately after you have your blog.

> **Take the Next Step**
>
> *Execute Tactic 5.4: Create Branded Content to Capture and Nurture Leads*

Create Content to Convert Leads to Customers

So far, you have created content to get your audience's attention, capture their information, and also nurture them through value and education. It becomes a challenge to construct messages that will effectively move a potential audience through the real estate process (bottom of the funnel). Because, at this stage, people have to decide whom to hire! Therefore, to win and be the agent of choice, you and your erudition must influence people's emotions into making that final decision.

Consequently, understanding the management of emotional intelligence in gaining relationships will positively affect the sales conversion process. As you already know, every real estate process is almost like riding a roller coaster of emotions. Buying and selling real estate could be a nervous experience. Anticipating your potential customer's emotional stage sets you apart from other competing agents; therefore, understanding emotional intelligence. Have empathy; be wary of your customers' emotions through the process, and sympathize with them throughout each stage.

For example, dealing with a motivated seller that has a contingent contract for finding another home is entirely different from a seller with no restrictions. The anxious motivated seller might be stressed, worried, nervous, and apprehensive. So you are not only required to maintain your emotions dealing with that but also have to work hard managing the seller's feelings. You might have to give your sellers peace of mind by providing evidence that will show your three components of credibility: competence, character, and care.

One of the most critical pieces of content that will help you move people from being leads to customers is mastering your sales scripts. Whether you are speaking to a potential customer over the phone, online, or in-person; you will gain more confidence and earn customers' trust by mastering your craft. By asking the right questions at the right time, you will move hot leads through the conversion process to become customers and win your marketplace. Practice, practice, practice! Is probably the most critical advice that I can give you in this chapter.

> **Take the Next Step**
>
> *Execute Tactic 5.5: Create Branded Content to Convert Leads to Customers*

Create Content to Convert Customers to Advocates

Go beyond conversion to grow and dominate your marketplace by consistently following up with your current database. Build content that will strengthen relationships not only with existing customers, but also with partners, affiliates, and vendors. Maximize your referral sources from people you already know - whether they are past customers or not. Stay in touch with your entire database via workflows and marketing automation campaigns. Scale your business by developing annual keep-in-touch messages. I call this "Top of Mind" content strategy.

The top-of-mind strategy means you're constantly "On."

Metaphorically, your store is always open for newcomers. You should keep open communication and touch base regularly. Turn the notion of ABC "Always Be Closing" aggressive sales approach into ABC "Always Be Communicating" and watch your business flourish. Begin measuring your success by the number of meaningful relationships you build, rather than the number of deals you close.

Nevertheless, whatever campaign you launch, remember, you are most effective if you pay attention to your audience's emotional stages. For example, you can ask for referrals after you close the sale since customers tend to be in the "happiness" emotional stage. Similarly, you can ask for a review or a testimonial when they're all settled in as customers tend to feel relieved, the "relief" emotional stage.

> *Take the Next Step*
>
> *Execute Tactic 5.6: Create Branded Content to Convert Customers to Advocates*

If you prefer professional "Done-For-You" content, use third-party resources such as *keepingcurrentmatters* or *rightlywritten* to develop your brand messages.

> *Take the Next Step*
>
> *Execute Tactic 5.7: Re-Purpose Branded Guides*

Top real estate agents learn from practice. They generate good content, know their stats, learn from their mistakes, practice, practice, practice, improve, then, they rinse and repeat.

Initially, you're neither going to create flawless content nor professionally produced videos. However, you are going to develop acceptable content to drive your real estate business upwards. Proceeding, your objective would be to find your brand style and the rhythm that connects you with your potential clients. Hence, preparing you for the next step of converting relationships to customers!

Organize Your Digital Content

One of the issues most professionals face is time wasted, looking for digital resources. Have you ever been frustrated looking for a document on your computer among a barrage of other materials?

Branded contents are valuable intellectual assets for your brand. Your research, time, and money invested in creating your real estate content will serve your business well if used frequently and consistently. The better you manage your digital assets, the less time you will spend looking for resources. Also, readily accessing well-organized digital content will result in optimized distribution channels. For example, when you can easily share a specific video to social media or a particular guide with a potential seller!

Your ability to search for digital resources by name, type, and the location is paramount. One may argue that all contracts, property photos, flyers, branded PDF contents, and videos are accurately filed on his or her computer. Managing documents offline is a risky business. You will be very disappointed if you lose digital content worth thousands of dollars because your laptop overheated and crashed. If you aren't backing up and organizing your documents online, do so immediately.

Take the Next Step

Execute Tactic 5.8: Organize Your Branded Content and Resources

PART IV

SCIENCE

CHAPTER SIX

SHARPEN YOUR AXE

"Give me six hours to chop down a tree, and I will spend the first four
sharpening the axe."
—Abraham Lincoln

What would any business be left with these days if we take away the web and technology? Not much, right? Online presence and digital technologies have been the crux of every company for decades. The more prominent your business is online, the broader your reach, and the more likely your business will thrive. The stronger the foundation of the technology, the more robust, advanced, and innovative the brand will be, and the larger it can scale.

In this chapter, I will introduce you to a plethora of web technologies aimed at strengthening the core capabilities of your real estate company. When appropriately applied, they will drive your brand ahead of your competition. Watch your business skyrocket once you implement these technologies along with the tactics shared in this book. Remember, success is a journey; don't try to integrate everything overnight. Start with techniques that make the most sense to you and your business, then scale and add more technologies as you grow.

Note: This chapter is slightly technical yet; I have tried to avoid technical jargon and use layman terms wherever possible. If you come across a situation where you can't implement a technology, please ask for professional help. Either a technical person in your office, someone you know, a marketing company, or outsource through a freelancing platform such as Freelancer.com.

Sharpen Online Presence

People like to work with people they know, trust, and like. Yes, in some cases, people prefer to work with known brands, but not as much when it comes to buying and selling homes. The real estate industry is a people-driven business. It is heavily reliant on relationships, trust, and credibility.

People don't order real estate agents off Amazon to show up at their doorstep and serve their needs. People work with agents they know and trust. They work with agents that understand their needs, wants, and desires. Thus, it would be best if you build confidence and worthiness online. In the real estate business, your personal brand is far more powerful than a business brand. Do you think Sally the Seller would work with an agent that she doesn't know just because Rose works for BHHS or Keller Williams? Certainly not! Though it is essential to be represented by a credible real estate firm, the human handling the transaction is far more critical to the client than the company.

That said, how do you build a strong personal brand online? Do you recall the forth chapter where you had to write your tagline, elevator pitch, unique value proposition, storyline, biography, and your executive summary? Well, now is the time to showcase that information online.

Show your character and who you are to those who don't know you and are surfing the web. Potential clients will feel more comfortable working with transparent agents who reveal their character online. Give your business a personal touch, and tell your story, and the anecdotes of other clients with whom you've worked. Let people into your life, show them that you care, and that you are undoubtedly different than most agents.

Set up Your Websites

There are three major types of websites that you must own: a blog, a business website, and landing pages. You could host all three on one single site or three separate websites. Nevertheless, they shall serve three different purposes. Build a personal blog to produce valuable content and strengthen your brand awareness. Also, create a business website with an Internet Data Exchange (IDX) integration. This way, you can serve potential buyers and sellers' needs. Finally, develop landing pages to capture quality leads and nurture your existing database into loyal customers. These types of websites are the most valuable digital assets for your company. They are the engine that repels your business forward.

Other types of websites that contribute to the success of your business online are not as consequential and significant as your sites. Because you do not own them, neither have full control over them. Someone else does! Think Zillow, Trulia, Realtor.com, Facebook, LinkedIn, and many others.

The biggest mistake most people make is investing far more time and effort into websites they don't own instead of the ones they do. However, I am not saying you shouldn't spend money with Zillow, but you should invest more money on your websites. Also, I am not saying you shouldn't engage with people on social media, but you should delegate more time on your websites. For example, syndicate your listings to your business website before sites like Zillow. Take the time to write a valuable article on your blog, and then share it a day later to social media. Doing so tells search engines that your sites are the authoritative owners of the content, neither Zillow nor social media.

Examine your social media profiles! Are you sharing content straight from your website and blog, or from external sources? Do you know what happened to all the content and time people invested in MySpace? Don't risk the chance of losing years of your time, investment, and hard work! Control your online destiny by investing in your websites and digital assets online!

Personal Blog

The most important investment you can make in your business is to start a blog! Real estate agents who started a blog a few years ago relied heavily and only on their blog in lead generation. They don't pay Zillow or Trulia, neither do they pay Facebook or Instagram, nor Google, Bing, LinkedIn, or anyone else. They create great content that generates quality seller, and buyer leads, consistently, all year long, every year!

Tell stories to get your message across using storytelling hooks that I revealed in previous chapters. Elaborate on topics you write about in long articles, preferably two thousand words in length and more. Additionally, use lead magnets and compelling calls-to-action on your blog to drive traffic to your landing pages and acquire quality leads.

Take the Next Step
Execute Tactic 6.0: Build a Blog

Business Website

The primary purpose of a business website is to serve the needs of a defined target audience. The Unique Value Proposition of the newest product is the first statement you read on a well-developed website's homepage. Then a call to action to dig more into values offered. For instance, Apple does a great job of serving its clients among other popular brands. When you visit apple.com, you will see their newest product on top, then two calls-to-action, either to buy or to learn more.

Your website must have a simple and professional design that is inviting so visitors can engage and stick around. The content and sales copy must be explicit so users can browse and easily find what they want. After all, your website must serve your business and your customers' needs. Here are seven key elements that your site must cover:

1. Help potential customers learn about you and your company. To learn about you, visitors usually look for an about us page, your biography or an executive summary.
2. Provide potential clients insight into your brand reputation using customer stories and testimonials. Anyone can go to Yelp for written reviews! Stand out from the crowd by adding a few video recommendations from your existing customers.
3. Allow website visitors to contact you by phone, a contact form, or a chat widget. To immediately respond to potential clients, make sure you receive instant text messages once they submit a request on your site.
4. Increase your brand awareness and online presence through Search Engine Optimized (SEO) pages. Hire a professional SEO marketer if your website isn't generating traffic from Google.
5. Direct potential customers to landing pages using lead magnets such as downloadable guides, white papers, market updates, CMA reports, and home valuation tools.
6. Track incoming traffic and on-page sessions to optimize and improve for better conversions.
7. Most importantly, help potential customers search home inventory using an IDX plugin.

Finally, make sure to include an about page, your bio, executive summary, UVP, success customer-stories, and testimonials on your website. Of course, add other helpful resources your target audience might find valuable.

> **Take the Next Step**
> *Execute Tactic 6.1: Build a Professional Website*

IDX Solution

Take full advantage of the advanced Multiple Listing Service (MLS) and Internet Data Exchange (IDX) search tools available in the market. Agent packages for as little as forty dollars per month can have a full-blown customizable home search and lead capture solutions on your website.

IDX tools are necessary for both buyers and sellers. Buyers surf the web seeking their next dream home, and sellers validate their listing information on the representing agent's website. Run a successful real estate business by satisfying the needs of both target demographics.

Provide an easy and effective way to search for homes, and look up featured listings on your website through a reliable IDX provider. Make sure you adhere to IDX rules and listing regulations. Display original listing company and agent names such as "Listed By." Also, disclaimers such as "Provided by Association Name," as well as the "Date and Time of Last IDX Feed Update."

> **Take the Next Step**
> *Execute Tactic 6.2: Integrate IDX into Your Professional Website*

Single Property Website

Imagine how thrilled your sellers will be when they see an entire website dedicated to their property. Separate property websites can be the key to get you more listings and impress potential homebuyers. They are essential for every home in your portfolio, not just luxury listings and high-end developments.

Having a single property website with the technologies that I am about to show you will raise the bar and set you apart from the competition. Making property sites as part of your business model will impress your sellers during and after each listing appointment.

A single property website could be a valuable element of your listing presentation. It will elevate the benefits to the sellers as it is used to market their individual property. Put together a single website for each home that is separate from your company's website. For nearly one hundred dollars per month, you can create unlimited individual property websites.

On each of these websites, you can install slideshows and virtual tours of the property. Add detailed descriptions to be found on the web, and other helpful tools such as mortgage calculators where buyers can calculate whether they can afford the mortgage payment. A single website can show nearby amenities, schools, local demographics, and an aerial street view. Best of all, individual property websites are responsive and can be accessed by computers, cell phones, and mobile devices—anywhere, anytime.

Interested buyers can contact you using the "chat" feature that will notify your cell phone of the inquiry so you can text them back. You can also receive emails of any requests made on the site. This way, you can get back to potential buyers with specific property information, track the performance of each single property website, and group incoming-leads into different segments. Powerful!

It is crucial to make the web link—also known as the Uniform Resource Locator (URL)—of each property website memorable and easy to type on mobile browsers—e.g., Safari or Chrome. This way, you can quickly respond to interested buyers with the URL when you receive text messages from the listing's lawn sign, a marketing material, or Zillow. Also, you can direct curious callers to the site for more information. When these interested buyers land on the site, not only will you collect their information to follow up, but also remarket to them using targeted ads.

> ### Take the Next Step
> *Execute Tactic 6.3: Partner with a Single Property Website Service Provider*

Landing Pages

You pay for a personal blog to build your brand. You create a beautiful business website to tell people about your outstanding real estate services. You pay for an IDX membership to fulfill the needs of potential buyers and sellers. How about capturing quality leads?

The mistake most marketers make with online websites—is attempting to land new opportunities while simultaneously trying to serve potential and existing clients. Your website, IDX search, and your blog aren't created to capture leads—at least not as the primary purpose. Instead, they are developed mainly to build brand awareness, educate, and serve your target audience.

To capture quality leads, you need to build landing pages—also known as squeeze pages. These pages are strategically designed to serve multiple purposes, but the initial aim is to capture quality leads. With landing pages, you can grab your visitors' attention, evoke action, and build your email list quickly and effectively.

Landing pages can also be used to measure the effectiveness of marketing campaigns as well as to nurture existing prospects in the database. When people visit your websites from different sources, you will encounter a difficult challenge in tracking which origin had the best conversion ratio across multiple sites. You can solve this issue by directing all web visitors to one location—your top performing landing page. Make sure your website, IDX solution, blog, and your ten social media profiles direct users to one landing page—or several if you are running multiple marketing campaigns. You can determine landing pages that generate the most leads by installing tracking scripts which I will cover in this chapter.

> **Take the Next Step**
>
> *Execute Tactic 6.4: Build Landing Pages to Generate and Nurture Leads*

Mobile Applications

Most real estate professionals use mobile technology to send new house alerts, neighborhood reports, custom property reports, and financial information. REALTORS®, sellers, and buyers are all using mobile apps to navigate the real estate industry.

The immense growth in the use of mobile technology has changed many industries, including real estate, from searching for homes to finding property values, communication, and social media. Portable applications have become an integral part of our lives.

Consumers expect real estate agents to provide information instantly. They anticipate accessing data on their mobile devices from anywhere and at any time. To thrive, real estate professionals must make mobile-first technology their top priority.

With mobile apps, real estate agents can conveniently text or email valuable information to sellers and buyers, communicate more effectively, and deliver information in real-time. These applications are highly efficient as they save time and ease collaboration.

> **Take the Next Step**
>
> *Execute Tactic 6.5: Develop or Subscribe to a Real Estate Mobile Application*

Top Real Estate Websites

The majority of web traffic and new leads come from top real estate websites such as Zillow and Trulia. Therefore, you must leverage these external sites as they reveal part of your brand. Whether the data they display, customer reviews, or your contact information; visit them often and ensure their content aligns with your brand.

Furthermore, log into the back end of all real estate syndication websites to manage your listings. Enhance your data to get better exposure by completing all required fields. Make sure the information is accurate, and take the recommended steps to optimize your listings.

Help sellers sell their homes quickly by providing more exposure to these top real estate syndication websites. Show prospective clients visits history, saved statistics, and performance graphs at your listing appointments. This way, you get the two-fold benefit from working the back end of listings.

1. Show how well paid ads of your current clients are performing.
2. Impress your prospective clients at each listing appointment of how well you manage their listings on these top real estate websites.

Take the Next Step

Execute Tactic 6.6: Update Your Information on Top Real Estate Websites

Other Real Estate Websites

Although I am covering technology thoroughly, as the months and years go by, you should make it a habit to research, find, and try what is trending. Examine mobile and web technologies that will make each process of your real estate business better, smoother, and more accessible.

Take advantage of existing and upcoming innovative technologies to save you time, make you more money, and beat the competition. The faster you implement new technology, the more likely you'll be ahead of every other real estate agent in your territory. Be an early adopter and be on the lookout for different technologies in your marketplace—at all times.

You don't need to implement every single technology, but you should be at least aware of most, should your business require them. Top agents are knowledgeable and use a vast number of the techniques mentioned in this chapter. Here are some that you should consider implementing in your business:

- **Syndication**: Set up syndication to advertise your listings on giant websites like ListHub and Point2Homes besides your MLS website, Zillow, Trulia, and Realtor.com

- **Marketing Templates**: Set up branded and professional real estate templates for electronic and snail-mail such as "Just Sold" and "Just Listed" flyers. Check out CoreFact, QuantumDigital, or SproutLoud.

- **Market Reports**: Produce flyers and postcards. Create property list reports, and neighborhood reports with local information surrounding property such as schools and their scores, nearby restaurants with their ratings and pricing, local golf courses, parks, and shopping malls. Also, include information about your local market in other marketing pieces you set up. Look into ListReports or AgentMarketing.

- **Electronic Signatures**: Streamline the real estate process for both sellers and buyers with electronic signatures. Make it more convenient for your clients to sign contracts and documents with every transaction. Use DocuSign or HelloSign.

- **Content Marketing**: Done-for-you content marketing is a Must-Have! Stay abreast with the market and get up-to-date real estate content. Post personal articles daily to all social media sites to educate your audience and gain more following. Hire an in-house content marketer or get a hold of KeepingCurrentMatters.

- **Diverse Marketing**: Strengthen your relationship with other agents, so you have an advantage when it is time to make an offer or negotiate a transaction. Take advantage of introducing yourself and your listing to agents outside your farm so you can receive referrals and enhance your brand awareness. Proxio allows real estate agents to market their listings and the listings of other agents to over 650,000 agents, in 15 countries, and 19 languages.

- **Online Forms**: Engage your customers and get them to converse with you. Install contact forms and feedback forms on your websites using Google Forms or TypeForm. This way, you can get a better insight into your client's needs. Most agents in the real estate industry do not survey their customers. Be different and set yourself apart. Listen to your clients and act on what they have to say.

Don't stop here! Make use of every little piece of technology available to set you apart from other competing agents in your local area, even when innovation doesn't directly contribute to revenue. For example, many agents still use the same sign they used ten years ago with the same old photo and just a phone number. Unless they've been able to put it under a lamppost, the sign goes completely dark at night. How can you be innovative with a street sign or add some "dazzle" to a "For Sale" old-fashioned lawn sign?

Besides the "Single Property Website" link, you can add a "Solar Sign Light." Your sign will then light up when the sun goes down attracting the attention of surrounding neighbors and people who drive by occasionally. While this simple act doesn't necessarily generate a massive amount of leads during the night, it does make you look like you have higher standards than other agents in your area. It tends to leave emotional affect and a memorable moment in those who see it, which may lead to future business. Look up "Solar Real Estate Sign Light" sold by AgentLight on Amazon to get your signs to light up at night.

A 2017 report from the National Association of Realtors® found that 62% of all home buyers prefer their agent to send property info via text message. By limiting yourself to calls and emails only, you may be shutting out a large chunk of your potential market.

The California Association of Realtors® found that only 5% of Realtors communicate by text. Using text messages as part of your real estate marketing strategy gives you a significant competitive advantage over other Realtors.

Receive curbside leads from your sign riders sent to your phone. Use a professional "Text Message Sign Rider" service such as ProAgentSolutions, DriveBuyTech, or AgentText to text potential buyers listing information with photos, pricing, and additional information directly to their mobile phones.

Take the Next Step

Execute Tactic 6.7: Subscribe to Applicable Real Estate Technologies

---

Add-On Websites

Add-on websites are tools used to optimize your website, blog, and landing pages. Optimization and conversion platforms will help you run split tests and experiments on your web properties to personalize the user experience across multiple devices.

Lead optimization and conversion is an ongoing process that you must always be improving to increase incoming leads and close sales. Refine your lead generation tactics and improve conversion ratios with web add-ons. Integrate online marketing tools into your websites such as Alexa, SimilarWeb, Google Optimize, Optimizely, Clicky, SmartLook, OptinMonster, and LeadFeeder.

> **Take the Next Step**
> *Execute Tactic 6.8: Optimize Lead Generation and Conversion Ratios*

Sharpen Online Tracking

Many businesses, including real estate, suffer in online marketing for many reasons. The two most prominent and commonly known reasons are lead generation and lead conversion. With the proliferation of web and social media platforms, companies struggle in creating a unified solution across platforms, channels, and departments—primarily sales and marketing.

Tracking online traffic will help you better advertise, manage, and measure the success of your real estate business. When you install scripts across a set of different platforms, you can:

1. Track online traffic and get useful insight across the web, social media, and display networks.
2. Convert the highest amount of traffic into leads, then customers.

The more insight you have into your customer journey, the more personalized your brand message and services will be. As a business owner, you want to understand your audience's behavior online and offline, so you can answer to their specific needs. Without a well-integrated system, it will be tough to grow and scale your real estate business.

You want to be able to measure your metrics daily and understand where you stand. Knowing your numbers and whether you have achieved your target goals requires a full stack of integrated sales and marketing platforms. Dive in and take advantage of available technologies. Once appropriately united, you will create a robust system that will grow your business rapidly.

You need to set up applications to manage your productivity, lead generation, the nurturing process, conversion, and monetization. These platforms will become the engine that will drive your entire business forward. Invest the time and money to learn how each technology contributes to scaling your real estate career.

While this may seem overwhelming, it is not! Set up one platform at a time and get help when you need it. With your growth mindset, willingness to learn, and commitment, you are destined to succeed.

Web and Social

Think about this: currently, Google and YouTube control the internet; Facebook and Instagram control social media. Regardless of what people are searching for, they're procuring information from the web and social media.

To capture most online traffic—especially at an early stage in the research process—accurately tap into web and social media. Maximize lead generation by escalating your reach to Google, YouTube, Facebook, and Instagram—or any other trending applications incepted after the writing of this textbook.

Also, you can focus on alternative traffic sources such as Zillow, Trulia, Realtor.com, or others. However, by doing so, you might be a bit late in the game as over 90% of traffic starts with Google, YouTube, and Facebook. You want to market to potential customers early in the real estate journey. For instance, by the time you target them on Zillow, most of those leads may have already been in some communication with a few agents. Real estate professionals who tap into web and social platforms will captivate such audience before starting the home search on real estate sites such as Zillow or Trulia.

While phone conversations and face-to-face communication may result in promising opportunities, online and social media ads can generate interest and drive traffic to your websites. Use tactics such as targeted ads in your farm, remarketing campaigns to a similar custom audience of competing agents, and early-phase newsletters to score potential deals. Whatever the communication type is, you must be present in that initial period of the home buying and selling research to compete. That is how top real estate producers capture most of the market share. They pre-set the stage to generate leads and acquire customers—months, and sometimes years in advance using pixels.

A pixel—also called insight tag—is a short snippet of code that you can install on your website or add within your email marketing campaigns. Pixels are used to collect digital information about users and their online behaviors. When a user visits a site or opens an email, the tag saves a small file on the visitor's device. This file is commonly known as "cookie." Marketers use "cookies" to get the most out of the web and social media ad budgets by targeting users who have previously interacted with a specific brand or service. Have you noticed a banner ad following you as you browse the internet? If so, you've seen first-hand the power of pixels.

Web and social media pixels will bring back those web viewers who interacted with your websites but didn't convert. Use remarketing channels such as Google, YouTube, Facebook, and Instagram with retargeting ads to bring those web visitors back to your brand. This way, you can turn them from strangers to quality leads and then customers. Next, we will dig into each channel and set up your real estate business for success on both the web and social media!

> **Take the Next Step**
> *Execute Tactic 6.9: Register for a Lead Remarketing Software*

Google Tracking Pixel

Google Tag Manager is a management system created by Google to manage JavaScript and HTML tags. Tags are used for tracking and analytics on websites and mobile apps without editing your site code. Use Google Tag Manager to handle all your web, mobile, and social media tags (pixels) to save time and hassle.

Sign up and install the script provided by Google Tag Manager on every platform you have. Add the code to your business website, your IDX solution site, your landing pages, your mobile applications, and any other digital platform you own and control. Seek a professional to install the Google Tag Manager script if you don't know how, or if you don't have a web developer.

Next, install every script, pixel, or tag mentioned in the rest of this book inside Google Tag Manager. Doing so will save you a ton of time because if you skip this installation, you will need to place all scripts directly on all of your platforms.

Once the Tag Manager Script is set up, you will also need to install Google User Analytics. Google Analytics is a pixel that will let you see how users interact across multiple properties on your web and mobile domains. Google provides both "Tag Manager" and "Analytics" at no cost.

> **Take the Next Step**
> *Execute Tactic 6.10: Install Google Tag Manager and Google Analytics*

Google Organic Search

Do you want to rank on the first page of Google's search results? Great! Hire a web administrator or take a tour of Google's Webmaster website. Sign up with a Google Mail account (Gmail) and add your sites. Webmaster by Google is a free service and tremendously invaluable.

Improve the ranking of your websites on Google's search engine with Google Console. Its tools and reports will help you measure your site's search traffic and performance. Leverage these tools to fix ranking issues and make your website shine in Google's search results. Receive email alerts when Google identifies problems on your site. See which Uniform Resource Locators (URLs) are affected by these issues and tell Google when you have fixed them.

> **Take the Next Step**
> *Execute Tactic 6.11: Optimize Your Websites for Search Engine Optimization*

Google My Business

Create a free business profile on Google My Business to help you reach, attract, and engage with local customers across Google Search and Maps. It is free and easy-to-use.

While Google My Business is a tool for businesses, it doesn't have to be. You can register an account with your name to promote your real estate services. For example, "Sally Smith Real Estate Consulting," or "Ross Smith Realty."

Set up Google My Business so you can manage your reviews, add photos of current listings, post to Google search results, and receive messages directly from users searching Google. Also, Google My Business will help you measure your search insights to understand the demands of your local market. This free service is, by far, the best and the most effective way to get found on Google, and yet, it is highly underutilized.

> **Take the Next Step**
> *Execute Tactic 6.12: List Your Business on Google My Business*

Google Paid Ads

Drive online visits, receive more phone calls, and grow your business with Google Ads. Use paid search through Search Ads 360. Get the most out of your search campaigns and respond to an ever-changing real estate market in real-time and at scale.

Get in front of potential buyers and sellers when they search your name on Google Search and Maps. Only pay for results, such as clicks to your website or calls generated from your business listing.

With Google Ads, you can manage ads on Gmail, Google Search, Google Maps, YouTube, Google Networks, Search Network, and Display Network. That is almost the entire web! Log in to Google Ads and set up your advertising tag by clicking the three dots in the "Google Ads Tag" box. Also, set up your website and YouTube Custom Audience. Yes, there is a "Custom Audience" feature in Google Ads as well as Facebook.

Google usually gives away advertising credit to new members. If you don't have a Google Ad account, go to Google and search for "Google Ads Coupons" before registering. One hundred dollars in free advertising credit is also accessible from your Google My Business account. Take advantage of this offer should it exist.

AdWerx Paid Ads

Adwerx is a display advertising service platform for real estate agents. AdWerx identifies anyone who does a real estate related search in your area. They follow them with your targeted ads on their Facebook, news websites, and around the web. People will eventually click on your ad, which takes them to your site and then to your featured properties. Because they keep seeing your ad, your name stays top of mind. Thus, you get people to see you everywhere, and eventually, connect with you.

Uncover new buyers and sellers with AdWerx. Find a new audience you aren't reaching today by targeting prospects in your local zip codes. Promote listings in your current territory and expand your business into new areas. Quickly create customized ads in minutes without the need for a designer. Start a display ad campaign to follow local prospects who visited real estate websites, looked at mortgage rates, or researched moving companies.

> **Take the Next Step**
> *Execute Tactic 6.13: Set up Your Business for Display Ads*

YouTube

YouTube is where the world goes to watch videos. More than a billion users—or nearly a third of all the people on the internet—are watching more videos on YouTube than ever. To be exact, and according to YouTube, two billion viewers each month seek out and engage with content that feels tailored to them.

As a real estate professional, YouTube provides you with a tremendous opportunity to connect with people on a personal level. Through video, you can reach people in the moments that matter—that is when they're looking for answers to their most pressing real estate questions.

YouTube reaches more people who are 18 to 49 year-olds than any television or cable network in the U.S. According to Google, in 2018, the average viewing session on a mobile device is more than 40 minutes long. In today's mobile-first world, YouTube brings video from primetime to all the time. It doesn't only entertain people; it enables them to learn anything from guidance, education, entertainment, and even real estate.

Do yourself and your business a huge favor and become a YouTuber; even better—a Vlogger. Turn every written article on your blog into a video. If this is not you, invest in producing high-quality professional videos to use for YouTube ads.

While you are still logged into Google Ads, log in to YouTube, and set up a YouTube Channel. It is free and easy to do! Once you're logged in, link YouTube to your Google Ads account. You don't need a video to do this. All you need is a computer or a phone and internet.

Now, you can use paid retargeting YouTube Ads to direct your custom audience to your blog, websites, and landing pages from YouTube.

Take the Next Step
Execute Tactic 6.14: Set up Your Business on YouTube

Social Media

Social media will soon become what I call "Reality Media." If you haven't fully emerged and tapped into social media, you are doing yourself and your real estate business a disservice. Social media has changed people's lives and revolutionized the culture of how we operate our business online. Also, it will emerge far more than we are currently witnessing. Hence, here are some thoughts on how I believe social media will be evolving.

Augmented Reality (AR) has already become a buzzword in the world of technology. Augmented reality is an interactive experience of a real-world environment where the objects that reside in the real-world are "augmented" by computer-generated perceptual information. It adds digital elements to a live view often by using the camera on a smartphone or tablet. That augmented view is where you are going to be able to insert your real estate ads.

AR makes advertising interactive, allowing marketers and advertisers to reach out to consumers in totally new ways. Many companies have already embraced this cutting-edge technology. Since technology is emerging at an incredible pace, you will soon hear about Augmented Reality Advertisement (AR ads) everywhere.

Thoroughly prepare for the technology of "Reality Media" so you and your business can be ahead of the curve. Begin taking advantage of social media platforms such as YouTube, LinkedIn, Facebook, Facebook Messenger, and Instagram for your video ads. Produce live video content on social media and be on the lookout for AR! Doing so will positively influence your real estate business and your lifestyle.

Facebook Ads

According to Facebook data in January 2018, there are nearly 1.4 billion people that use Facebook every day. Capture the attention and prompt action of millions of people using Facebook Ads. Create cookies (pixels) where you can retarget viewers, saving marketing dollars by being more specific about who sees your marketing ads.

Proactively reach new customers and retarget existing people in your database through Facebook ads. Facebook marketing is, by far, one of the most effective lead generation tools that exist online.

The first step to creating effective Facebook ad campaigns is by building a Custom Audience. Second, write a compelling ad copy— which I covered in the previous chapter. By mastering these two aspects of your overall Facebook marketing strategy, you will be ahead of your competition, and you will generate more leads than any other online platform.

Boosting a post isn't going to get you leads. You have to dig into Facebook Ad Manager, write useful ad copy, test different ads, designs, placements, and formats. For example, through testing, I've learned that creating tons of videos and promoting them on Facebook will result in the cheapest cost per impressions in comparison to text content and posts. Facebook loves videos—to encourage video, Facebook made video ads the cheapest.

Marketing to a particular type of audience is far more rewarding than a "boost and see if it sticks" kind of ad. For instance, create seller-targeted ads, and consequently, FSBOs only ads, and Expireds only ads.

Besides, a professional video ad is far more effective than a written advertisement! Ads that tell a story about a past client's experience or start with a customer testimonial or some social proof are the most effective. Test what works for you and your farm, and then double down on what works best.

Install Facebook Pixel

Install your Facebook pixel everywhere by adding the Facebook script tag into your Google Tag Manager. Visit Facebook for Business to see instructions on how to install the Facebook pixel, hire a professional, or search YouTube for "how to install the Facebook pixel."

To effectively use Facebook Ads and remarket to people who have previously visited your brand, you must install the Facebook pixel on all your web pages. Install the pixel on your business website, IDX website, blog, and landing pages through Google Tag Manager. The Facebook pixel will contribute to generating more leads by redirecting paid traffic back to your brand.

Upload Your Custom Audience

Hyper-target your audience for highly relevant, personalized, and enticing ads! Carefully build specific groups of segmented customers or audience to improve brand awareness, customer engagement, and maximize your return on marketing investment.

Upload your existing database to Facebook Custom Audience. Each contact must either contain a mobile phone number or an email address. The rest of the demographic information is optional. However, the more fields you upload, the more targeted your ads will be.

> *Take the Next Step*
> *Execute Tactic 6.15: Advertise Your Business on Facebook*

Facebook Messenger and Chatbots

Have one-on-one conversations with your local farm through Facebook Messenger. Utilize it to communicate with brand new prospects and loyal customers. Messenger allows you to drive awareness, acquire new clients, and support existing ones—all in one continuous conversation.

People are increasingly shying away from social media feeds and turning to messaging applications to connect with new and existing businesses. Brands are also increasing confidence and credibility with potential clients using Messenger due to its high engagement rates and instant support. Messenger can help you acquire more customers by providing a direct, conversational way for people to take action where they already spend most of their time.

Connect with existing customers to enable transactional support via Facebook Messenger. Private messaging increases transparency between you, your team, and your clients. Integrating Messenger into a client service strategy allows customers to communicate their real estate needs with your business on their terms. No more waiting on hold or keeping website windows open. The conversation stays in Messenger, which is easily accessible for questions and re-engagement at any time and from anywhere.

Facebook Messenger isn't just for building presence and driving awareness. You can use it to show your customers that you care. When you find yourself sending the same message repeatedly, you can take advantage of chatbots, so you're always available. For example, you can automate information on a specific listing with a chatbot. Also, set up automated messages to greet your audience and potential customers who message you after business hours.

> *Take the Next Step*
> *Execute Tactic 6.16: Strengthen Relationships with Private Messaging*

Facebook Audience Network

All big brands use display ads because they are mighty and remunerative, and so should you. Have you ever looked at a product online and then noticed advertisements from that company are following you around the web including Facebook? Whether you find it "creepy" or "cool," these types of ads are lucrative—if done well and frequently. This phenomenon is called remarketing through Display Network. It is the same as Display Network Ads used with Google Ads.

Leverage the power of Facebook ads. They work because they're relevant for people and uncomplicated to create. Besides, ad reports make it easy for any business to measure the success of each ad campaign. Facebook's Audience Network brings these same powerful features to advertisements on apps and sites beyond Facebook, where most people spend their time. You can deliver Ads within the Audience Network across devices into a variety of videos and display placements, including native, interstitial, rewarded, and in-stream videos. Taking advantage of Facebook's Audience Network can be extremely powerful, especially if you have a real estate mobile application.

Regardless of the placement type, you don't need to create and upload new creative assets. You can create one ad in which will automatically render to fit the placement type with the same creative assets used on Facebook. Though, you always have the option to customize your creation. The most effective ads are usually videos. Use the same professional video you've created for YouTube ads with Facebook Audience Network.

Take advantage of Facebook's Audience Network and learn how to customize your creative assets on the Facebook Advertiser Help Center.

> **Take the Next Step**
>
> *Execute Tactic 6.17: Go Beyond In-App and Extend Your Reach to the Network*

Instagram Ads

Inspire action through visual expressions of your real estate brand with Instagram Stories and IGTV. As of this writing, Instagram Story Ads are the most lucrative advertisements on social media. They are new, trending, interactive, and exclusive. One caveat though, an Instagram account must have a minimum of ten thousand followers to qualify for the "Swipe Up" ad feature. Think about this; most real estate agents don't have ten thousand followers. If your account isn't eligible to advertise on Instagram Story yet, get on IGTV. Take advantage of Instagram Stories to promote your real estate services without the need to pay for ad space. Use hashtags, polls, and other Story tools to engage your audience. Also, verbally tell people to visit your website or call your office.

What if you don't have enough followers? That is common! There are many ways to get followers. While some agents may pay hundreds of dollars to get fake followers, I don't recommend it. Instead, you can automate much of Instagram activities to increase your following. You can do so by using services, such as FollowLiker or Phantombuster, to automatically like and comment on users posts via location, hashtags, and demographics. However, you must engage with people to get followers and drive consistent traffic to your profile.

Nevertheless, you can still use Instagram ads for marketing and retargeting your audience through the Instagram feed. Instagram has been one of the primary sources for lead generation for many real estate professionals. Further, Instagram ads have a better conversion ratio than Facebook. I have surveyed real estate agents myself! Also, I use both platforms. Instagram always outperforms Facebook. Besides, people come to Instagram to be inspired, explore their passions, and discover new things. Hence, posting pictures of open house events and videos of listed homes for sale in your area may attract newcomers.

> **Take the Next Step**
> *Execute Tactic 6.18: Advertise Your Business on Instagram*

LinkedIn Ads

LinkedIn can help you find highly targeted customers and partners. The targeting on LinkedIn is unparalleled in the realm of digital advertising. You can zero in on a specific industry, company size, and job role.

Upgrade your LinkedIn account to premium so you can use Sales Navigator, access in-depth insights, send mass messages (up to 50 connections), and use the InMail feature. Premium is free to try for 30 days. Also, sign up for LinkedIn Business to advertise and generate quality leads. Search Google for "LinkedIn fifty-dollar coupon for new advertisers" or create a new account and wait a few days before you run any ads. LinkedIn might send you a fifty dollars coupon via email.

Both Premium and LinkedIn advertising can help you grow your connections, increase brand awareness, generate more leads, and establish partnerships. Imagine the possibilities of using Sales Navigator to connect with local builders, attorneys, accountants, and bookkeepers to get referrals. Set your advertising campaigns with targeted ads to be displayed to high-end audiences only such as doctors, engineers, founders, and C-Level executives in your local farm. That is what top real estate professionals do—target vast opportunities for higher returns.

Approach your prospecting efforts and marketing strategies based on specific professions with high net worth. This tactic is worthwhile due to your high relevant messaging about the customer's industry. Remember, the most significant two moving factors to a successful marketing campaign is getting your audience's attention and writing a concise brand message. LinkedIn allows you to do just that.

> **Take the Next Step**
> *Execute Tactic 6.19: Advertise Your Business on LinkedIn*

Social Media Groups

Generating quality leads from social media groups is by far, one of the best organic—not paid—tactics that I have experienced in the last decade.

Create a real estate group on whichever platform you use often and are most comfortable with, either Facebook or LinkedIn. Then invite your family members, friends, sphere of influence, top partners, and past clients to join your group.

Search for and join other groups created by someone else in your community. For example, find groups formed by your city, schools, homeowners associations, or pet stores. Your customers tend to hang out in these communities.

Join at least 3-5 groups and become the awesome human in the comments. Follow the 10/1 rule: For every one post on your social media, comment and engage ten times on other people's posts. Look for those with active engagements—that is, a high number of comments—and help others answer their questions in the comments section. Members will eventually notice you and visit your Facebook or LinkedIn page. Then, lead them to join your group so you can prospect them, build relationships, and turn them to clients.

Find prospects by engaging with the members of the groups you have joined. Once you're inside the group and approved as a member, find active ideal candidates, and then invite them to join your group by sending personalized invitations. Once these invitees join, you have all your best prospects in one place!

Your social media group has now become a sacred place. It will allow you and your tribe to build stronger relationships. You can control your group so that no competitors or other real estate agents get in.

You need to strengthen your credibility and build trust by sharing valuable content that your prospects will love. The guides and videos you've created in earlier chapters are great examples. Now, you can put them into use. You also get to demonstrate your values and expertise to them while avoiding overt sales pitches or SPAM.

Besides, don't make your group about you. Social media groups are about your members, potential buyers, and sellers in your community. Plus, you have a built-in email list, a focus group for your core prospects, and social fans. Therefore, use social media groups to build your brand and generate leads to boost your real estate business.

Take the Next Step
Execute Tactic 6.20: Build Loyal Tribes with Social Media Groups

Social Media Management Platforms

Save time and start doing more with social media by using an all-in-one social media management platform. Manage all your social media accounts in one place. Schedule content and check analytics for all your social accounts, rather than going to eight different networks. Schedule social media content to post to all your profiles at once and at the ideal times your fans are online. Platforms such as Buffer and Hootsuite will publish them automatically, according to the posting schedule you put in place.

Review your social media analytics to track how your posts are performing. Measure the impact of your social media campaigns through comprehensive reporting. Measure conversions by social media channel, then separate return on investment (ROI) between owned (free) and paid media. Test what works and what doesn't. Analyze such data to improve your results and increase engagement.

> *Take the Next Step*
> *Execute Tactic 6.21: Manage Your Social Media Presence*

Sharpen Online Management

There are two sides to your real estate business: the face of your brand, and the back office. Now that you have set up your web and social media platforms, you need to set up applications to manage what is behind-the-scenes. It would help if you had more technologies to facilitate your daily activities, business processes, incoming leads, current customers, and the overall business.

Managing your real estate company with sticky notes and notebooks won't cut it. If you want to excel and dominate your marketplace, most of the platforms mentioned in this chapter are as critical as having a bank account. To establish a reliable lead generation and conversion system, you need to streamline your lead management, automation, and collaboration platforms. Each plays a role in optimizing your real estate business and contributing towards your success.

If you feel overwhelmed with the number of technologies, platforms, and mobile applications you must manage, you are just at the tip of the iceberg. Dive in, and dive fast! Technology is growing at a much faster pace than we humans can begin to fathom. It would be best if you adapted swiftly and intelligently. The sooner you embrace technology, the faster you will excel and dominate.

Contact Relationship Manager (CRM)

A contact relationship manager (CRM) is a MUST-HAVE for any business. It helps to manage the database, nurture existing leads, follow up on opportunities, engage with clients, and improve productivity.

Add the contact information of every person you come across in your life to your CRM. Then, follow up with them to turn them into happy customers or referring partners.

Manage your contacts with ease, qualify people, set reminders, set up automated drip campaigns, and scale your real estate business with one of the CRMs mentioned in the next tactic.

> ### Take the Next Step
> *Execute Tactic 6.22: Purchase a Cloud-Based Contact Relationship Manager (CRM)*

CRM Tools

Now that you've added every contact within your network to your CRM, prospect and follow up daily for at least two hours. Implement a consistent communication strategy to stay top of mind with your prospective leads, potential opportunities, and pending deals. Moving candidates through to closing is more about taking action by following up. Make sure you put in those two hours daily, working your database, and managing your CRM. For example, follow up from 9 AM to 10 AM and 3 PM to 4 PM.

Integrate your CRM with other online tools to streamline the process and scale closing capabilities. For example, connect your CRM with a platform named Remine. Remine can feed your database with potential buyers and sellers. Also, link your CRM to BombBomb so you can auto-respond immediately to incoming leads with personalized videos, to name a couple.

Sometimes agents find themselves sporadically hosting their leads in multiple places. For example, they save them in spreadsheets, several online lead providers, or a previously used CRM. Use a cross-platform syncing mechanism or software to bridge your leads and clients' information in one system.

In other cases, you might want to also sync your contacts and sphere-of-influence (SOI) across multiple platforms and devices, such as your CRM, your phones, and a backend database. For instance, use a contact syncing software to unify the data in your database with the contact information of people and clients saved on your mobile phone.

Take advantage of all available CRM tools to combine and sync your entire network, leads, and contacts. Facilitate a seamless CRM automation process with cloud syncing capabilities. Streamline a large part of the real estate process, and uncover a powerful potential in increasing your customer base, you didn't realize your business had.

Take the Next Step

Execute Tactic 6.23: Incorporate Seamless CRM Automation with Cloud Syncing Capabilities

Business Tools

From web analytics to optimization, conversion, and retargeting, now, you'll be able to manage lead generation, acquisition, and conversion from end-to-end through digital marketing platforms and sales management tools.

Go beyond the closing stage to survey your customers and get their feedback. Show your clients that you care by asking them about their experience and what you can do to improve. Hearing what they have to say and acting on that can enhance your business process and your services. Besides, you can leverage surveys as another great way to ask for future referrals.

Take the Next Step

Execute Tactic 6.24: Streamline Your Real Estate Business Digitally

Marketing Automation

All the tools and platforms shared with you so far were to set up the foundation, acquire, and manage your business. How about the mechanisms that drive growth and revenue to your real estate business?

Automation Platforms

Instead of signing up for many sales and marketing platforms, sometimes, if budget permits, it would make much more sense to sign up for an all-in-one Marketing Automation Platform, or MAP.

MAPs provide tools for blogging, SEO, social media, landing pages, CRM, drip campaigns, email, text, web analytics, and more. While large teams commonly use marketing automation tools, it doesn't mean you can't run your business like one.

Notes

Sometimes you come across situations where you need to make a quick note, but you don't want to insert it into your CRM immediately. Why not ditch the old-fashioned notebook and use an online journal that feeds directly into your CRM when needed? For example, integrate Realvolve CRM with Evernote or KVCore CRM with Google Keep. This way you can refer to your notes during listing appointments or when you're out in the field.

Appointment Scheduling

Nothing is more frustrating than missing an appointment with a potential client or being late to one. Besides using a cloud calendar to set up your arrangements, there is another option that allows others—including clients—to book time with you which syncs right into your schedule.

Organize your entire business with a robust online calendar that's fully integrated with your company and daily plans—schedule meetings without the back-and-forth emails with services like Acuity Scheduling or Calendly.

Phone Dialer

Some agents love it, and most hate it. Phone prospecting is bittersweet. If you choose to work FSBOs and Expireds, then research ArchAgent, Mojo, Kixie, and SalesDialers—a few phone dialer service providers.

Call Tracking and Recording

Having a web application that automates the tracking, recording, and engagement of inbound phone calls across all marketing channels is vital to your business. How can you tell if a phone call lead came from a social media ad or a street sign? Asking each potential client isn't a reliable metric in the long hall. You need to be able to run reports to measure which marketing channel and efforts are bringing in the most phone calls and the best quality leads. Use call tracking and recording services such as CallAction, CallRail, DialPad, or Vumber to record incoming phone calls and track the performance of your marketing and sales campaigns.

Text Messaging

Send new listing alerts, Open House invites, or price drop notifications by using SMS to capture leads and grow your database. Track outgoing and incoming text messages. Send or schedule group text message campaigns. However, if you want to avoid the footer verbiage, "Reply STOP to Unsubscribe" send a maximum of 20 text messages at once.

Before signing up for a text messaging platform, investigate available CRMs that already provide this feature as a built-in function. For instance, Chime, KVCore, Realvolve, LionDesk, and many other CRMs are equipped to send SMS. Besides, you can look into Google Voice, EZTexting, SimpleTexting, SlickText, SendHub, TextMarks, and TextMagic for standalone text messaging service.

If you are a large office and want to send text messages in mass, you will need to use US Short Codes. Sign up for your Short Code through US Short Codes or Twilio's professional services.

When sending SMS marketing campaigns, personalize your text messages, and keep them short and concise. Here are a few examples:

- *Quick question: Are you still looking to buy a home soon?*
- *Are you thinking about selling your house this year?*
- *Want to grab coffee this week? On me!*
- *Just Listed: 4 Beds, 3 Baths. www.bitly.com/123MainSt*
- *You're Invited! Open House: 123 Main St. Sat. 2-6 PM.*

Voicemail Drop

Quickly and efficiently use the voicemail broadcasting service such as Sly Broadcast, Sly Dial, Dial My Calls, or Trumpia to send out automated voice messages to a list of phone numbers. Since you can't text landlines, leverage ringless voicemail to broadcast in bulk, schedule in batches, or send immediately to both landline and mobile phone numbers.

Check your phone list with data verification services such as informatica or data247 to help you and your real estate organization stay compliant with USA Federal Do-Not-Call (DNC) list laws.

Disclaimer

I am not an attorney, but it is illegal to text people who didn't subscribe to receive your marketing messages. Refrain from sending text messages to prospects that didn't opt-in to promotional content. When doing so, phone service providers might block sending and receiving your text messages. Also, companies that illegally call numbers on the National Do Not Call Registry or place an illegal robocall can currently be fined up to $42,530 per call. Thus, send text messages and voicemail drops to subscribers only.

Chat & Chatbots

Install a professional chat plugin such as Intercom or Drift on all your websites. It would also help to install an AI-Powered Chatbot such as Instabot or Mobile Monkey. Chat with clients online or have an Artificial Intelligence short messaging system automatically qualify and respond to your web visitors. Nonetheless, make sure you connect incoming leads from the chat widget with your CRM system so you can seamlessly follow up with them.

Email

Rumors have been going around for many years that email will be dying down. It is a rumor! Email is here to stay and will certainly not be going away anytime soon—at least, not for a while. Email marketing automation has been—for many agents and real estate companies—the number one driver for revenue.

Pick an emal platform that fits your needs and your budget, and then stick to it. While there might be a slight difference among all email marketing software, they all have similar features, benefits, and functionalities. Try MailChimp, Benchmark, AWeber, ActiveCampaign, GetResponse, or BombBomb, and use the one that you find most intuitive.

Push Notifications

Have you come across a website that slides in a notification box, giving you the option to "Allow" notifications or share your location? Well, that is how you subscribe or deny subscription to the browser's web push notifications on your desktop or laptop.

While answering emails or browsing the web, a small notification box pops up from the bottom right corner of your computer, "Your Facebook friend just posted a photo." Then you click! That's a perfect example of how you can send your potential customers push notifications to bring them back to your website.

Push notifications come in two different ingredients, web (desktops or laptops), and mobile. While it is easier to collect web subscribers for desktop notifications as potential customers visit your websites, it is much harder to obtain mobile subscribers. You will need either a mobile application for people to download or a beacon installed in a specific location so you can accumulate and manage subscribers.

Not only will users have to subscribe via their mobile device, but iPhone users must enable notifications through their device Settings to receive them. According to a study by Statista, there are 63% iPhone users in the US versus 37% who use Android. Despite the complex process of mobile subscription, it is well worth it. Use OneSignal, PushCrew, or SendPulse to start taking advantage of web and mobile push notifications.

Video

Make videos. A lot of them! The more videos you produce, the more people you'll likely reach. The more people you approach, the stronger your brand awareness, and the more likely you'll succeed in closing more transactions.

People are currently consuming videos more than any other communication channel—e.g., YouTube, Facebook Live, and IGTV. Give people what they want! Impress them with your professional-looking videos and presentations! Create explainer videos, advertising videos, and listing presentation videos. How impressive would it be to walk into a listing appointment paperless, with just a tablet and a video highlighting your values and marketing plan? Here's a great idea; send the video to your seller before your presentation to separate yourself from every other competing agent in your market. Doing so will elevate your values with your potential clients instantly.

There is no limit to what you can do with videos. Your possibilities are infinite and limitless. You can welcome new clients, educate the community, explain the real estate process, request feedback and testimonials, thank sellers and buyers of newly closed sales, interview local shops, promote your neighborhood, and feature local attractions through video. For example, search YouTube for "Welcome to Montclair, NJ, with Nick & Anne Baldwin." See how Nick & Anne featured their Montclair hometown attracting new buyers from all around New Jersey. Besides, video emails build trust, convert leads, and win more referrals. Get personal and make more videos to acquire more face-to-face appointments. Videos do work. If you haven't been producing video content, start immediately! Take advantage of services like BombBomb, WeVideo, Wistia, Vidyard, VideoBlocks, and Animoto to create professional and authentic high-quality videos. Use video to capture your audience's attention and convert them to customers.

Take the Next Step
Execute Tactic 6.25: Automate Real Estate Sales and Marketing

Sharpen Online Collaboration

How do you manage all the above tools and technologies, and still run your real estate business like a champion? Easy! By signing up for a few more platforms that will manage them all, help you collaborate, and increase your productivity.

First, you need to easily and quickly map your brain, jot down any new ideas that come to mind, and manage existing ones. Use EverNote, Workflowy, TaskRabbit, or ZipLogix to keep yourself and your thoughts always organized.

Next, you need a Cloud Storage Service to document and easily find your digital assets. Stay focused by organizing your notes, documents, videos, and any other digital content you own. Don't be the chaotic real estate agent that spends hours looking for a vital record or the person that saves everything under the sun to the desktop. Keep every single paper backed up into the Cloud. Don't risk misplacing documents or losing data. Be organized and store your digital assets online using Dropbox, Box, SugarSync, or Google Drive so you can manage your business better.

Third, you need an ideal and straightforward project management platform. Top real estate producers use this daily and frequently. How else would you be able to manage your successful business and still scale it without affecting productivity? Project Management Platforms are designed to maximize your time, your teams' productivity, enhance communication, and improve the overall customer service.

Organize and manage projects, teams, or departments. Manage websites, designs, content, clients, budgets, and more all in one place. Ease your online collaboration and increase productivity by tracking tasks, due dates, employees, and clients.

Stop the frustrating phone calls and emailing employees back and forth. Collaborate more efficiently with groups of people. Get full visibility into every aspect of your business. Create visual project plans to see how every step maps out over time. Pinpoint risks and stay in complete control of your real estate business by using a project management platform such as Trello, BaseCamp, or Asana.

Finally, you want to be able to see your entire company's performance at a glance. Create an online dashboard that syncs across vendors and measures data across all platforms and devices. Having an eagle-eye view of every piece of live data and full access to insights within your real estate company is robust and highly effective in optimizing your business. Although dashboards are essential in providing quick access to data and insights, they are the least utilized in the real estate industry. Klipfolio, SuperMetrics, and PowerMyAnalytics are worth considering.

··

: *Take the Next Step*
: *Execute Tactic 6.26: Sharpen Online Collaboration*
··

Build the System

I could create an entire book around this chapter only. Even then, I could not express how positively each platform and technology could impact your profession. This chapter was moderately lengthy, but it is vitally crucial that you make needed technologies part of your Standard Operating Procedure (SOP). Do not let the technical aspect of this chapter intimidate you! Invest the time to understand the benefits and how each tool can revolutionize the way you run your real estate business. There is a ton of information, and indeed, plenty of value.

Every technology in this chapter will contribute towards building your domination system and the overall revenue of your company. Leveraging technology to develop a growth engine is a must in any online business today, including real estate. Don't let the domination system be "You." You are not the system! You want to be able to monetize while you are on appointments, asleep, on vacation, or unavailable. You want to be flexible and free. You must build the system, not be the system!

Metaphorically, think of your business as an engine. Currently, the majority of real estate professionals act as fuel. Once they're removed or burned out, the power shuts down, the entire engine stops, and so does their business. Hence, over 80% of real estate agents nationwide enter the real estate industry, and exit in less than three years later with regrets. Those are the ones that didn't sharpen their axe, neither took full advantage of technology nor built a scalable system.

Having a reliable system that generates income without you having to wear different hats is a precious business. You need to be—not the fuel, but the person that adds fuel to the engine.

You are the inventor and the creator of your domination system. You are the skilled talent that builds, manages, and improves the momentum of the system. Again, you're not the fuel, neither the system. You are the one who produces it.

Keep in mind, most of the technologies highlighted in this chapter integrate seamlessly together. Shop around, find out which platform best suits your needs based on whether you are an individual real estate agent, a small team, or a large brokerage. Make your sales process and workflows more efficient by connecting the tools you use daily.

Create your smart real estate domination system! Learn, practice, excel, and continuously be perfecting your system. Domination is a process that you control. The more time and energy you put into it, the more it will give you!

Take the Next Step

Execute Tactic 6.27: Sharpen Your Axe

> *"If you overthink, you will not do; because you can do more than what you think. Get started now and think as you go!"*
>
> —James Tyler

CHAPTER SEVEN

Build Massive Momentum

"The future cannot be predicted, but futures can be invented."
—Dennis Gabor

What a coincidence! Rose the Realtor, launches Instagram, and she notices in the notification section *"Your friend Fred Smith is on Instagram as InstaFredSmith."* Then, when she opens the LinkedIn app similarly, Rose the Realtor notices a notification *"Your contact Fred Smith just joined LinkedIn."* What does she do? Without hesitation, Rose the Realtor follows her friend Fred Smith, connects with him, and sends him a quick message saying *"Hi Fred, how are you?"* Likewise, you would do the same, right? Well, the exact phenomenon happens on Facebook, Twitter, Pinterest, and Snapchat. If Fred joined those social media profiles, these applications would notify Rose to connect with him.

These occurrences are not coincidences! Social media platforms have access to all the contacts on our phones. They can see—meaning an algorithm can predict that Rose is friends with Fred. So, as soon as Fred joins a social media platform, Rose gets notified.

Now, let's assume Fred has been on Instagram, but Rose wasn't aware of that—or she didn't see the notification. Rose will eventually notice Fred on the feed in the "Suggestions for You" section, and pretty much in every corner of the app. Rose can then follow Fred and become friends on Instagram. Similarly, LinkedIn will suggest Fred as a "Connection" on Rose's LinkedIn account. Also, Facebook will recommend him as a friend in the "People You May Know" section, and so on. After all, they are friends in real life. Social media platforms recommend that they connect.

When we connect, follow, and socialize with people, we build relationships. Sometimes we develop great relationships with some people that become our repeat clients and favorite friends. Thanks to social media for making this possible. With our smartphones and social media, we can now connect and communicate with anyone, anywhere, anytime.

Hypothetically, what would happen if Rose adds the entire population of a specific city as contacts into her mobile phone? Yes, you guessed it. Everywhere Rose turns on social media; those apps will think Rose is a friend of every single person in that city, and hence, suggest she connects with them. While this is an excellent way for Rose to communicate with people in her town, adding an entire city's population into her phone will create massive chaos. Rose doesn't want people in a whole village suggested to her on social media. She only wants people who might be interested in her services—buying and selling real estate.

What if I told you there is a way for Rose to connect with potential homebuyers and sellers in any specific location? What if she can turn what could've been massive chaos into what can be tremendous momentum that will consistently scale her business with enormous growth?

The Momentum Concept

When I said, this part of the book was "science," I meant it. Let's discuss a physics equation: *Momentum = Mass x Velocity (p = mv)*

Linear momentum is consistent with the intuitive understanding of dynamic energy. A large, fast-moving object has a more significant force than a smaller and slower object. All objects have mass; if an object is moving, then it has momentum—it has mass in motion. The magnitude of an object depends on how much mass is in action and how fast it is moving. Thus, linear momentum is the product of a system's mass multiplied by its velocity: *Momentum = Mass x Velocity.*

Mass

By definition, mass is a group of people or objects crowded together." In your real estate business, where is the most significant number of people crowded together? Your database! Your email list is a database. A spreadsheet with contact information of people within your network is a database. The Contacts on your mobile device is also a database. Thus, it is safe to define mass as the total number of people in your database. To increase the momentum of your real estate business, you must increase the number of people in your database.

Consider this, which real estate agent would have a higher chance of closing ten potential clients consistently every month? Would it be Ross, with 200 contacts in his database, or Rose, with 6,000 contacts in her database?

Velocity

Velocity is the speed of "something" in a given direction. Obviously, in our equation, that "something" would be the number of people in your database. So, to build more momentum, you must increase the rate or speed at which you add people into your database. For example, who would have a higher chance of increasing the number of closings consistently every month? Ross, by adding one contact to his database daily (8-hour velocity rate), or Rose by adding eighty connections to her database daily (8-hour velocity rate)?

Imagine you are Rose adding eighty contacts to your database daily versus the average agent, Ross, who adds only one person to his database a day. What kind of momentum would your real estate business have in the next 90 days?

- You, or Rose: 80 x 90 = 7,200 people
- The average agent, or Ross: 1 x 90 = 90 people

The Three Momentum Drivers

Scientifically speaking, you build a higher momentum the faster you add more people to your database. However, we measure business performance by closed sales. We don't measure success by the number of contacts within our network! Hence, increasing momentum requires other tasks that will convert your database to customers.

When you look at the top producing agents in the nation, you'll notice that they all share one common ground. They all have a minimum of two teams. The marketing team is responsible for maintaining an optimum database and consistently growing its size with new contacts. The sales team is responsible for setting appointments and closing more sales.

While most think sales and marketing are two separate departments, they truly complement each other. Top producers understand that both sales and marketing teams must work hand in hand towards achieving one goal—growing the business.

There is one other department that is far more important than sales and marketing. Its primary role is building awareness and creating demand. Business Development (BD) is the core foundation of building any business. It is responsible for establishing relationships that will deliver growth. It combines sales and relationship management skills to develop new business opportunities. The BD department is often linked very closely to the marketing department as its tasks and efforts contribute to the best quality of leads passed onto marketing.

The business development department feeds the marketing department with potential leads—BD plants the seeds for the marketing department to nurture them. The marketing department nurtures those leads until they are ripe and ready for the sales department to close them. The sales department harvests qualified leads, turning them into customers. This cycle is what keeps the force going for any business. With a prime momentum comes a consistent flow of capital, and therefore a long-lasting, thriving business.

A business is as healthy as its sales cycle, but sales will have little to no success without BD and Marketing. A salesperson isn't typically responsible for strategizing the business model, neither the product-market-fit nor setting up marketing campaigns. It is imperative to keep these three departments in sync to serve all business and customer needs. A non-healthy and struggling business model is the one that expects the momentum cycle to keep going.

To run a successful business of any type, including real estate, it would be best if you understand the importance of these three departments. They must integrate seamlessly together as they are the main drivers of momentum, revenue, and your potential to succeed.

Therefore, to close more sales and increase the revenue of any business, you have to keep the momentum cycle going by building and executing onto these three departments:

1. **Business Development**: Build a massive database by adding the most significant number of people to your network at the highest speed possible. Plant more seeds using a direct outreach plan to create an ongoing demand and ignite enormous momentum.

2. **Marketing**: Generate more leads and nurture your existing network with valuable content.

3. **Sales**: Grow your real estate business by converting the highest number of people into customers at the highest rate possible.

As you can see, these three drivers of momentum are vital to the growth of your real estate business. In the next few chapters, you will learn how to become the Chief Executive Officer (CEO) of your business by mastering these three areas and leveraging every aspect of their functions. Let's begin with the crux of business development; building your database.

Build a Massive Database

Ross can't possibly close hundreds of listings and generate millions of dollars in sales each year with only two hundred people in his database. On the other hand, Rose has that potential due to the size of her massive database.

The smallest database size of a top producing real estate professional that I have seen was, on average, around 5000 people. If you want to be like Rose and compete with top producing agents in your area, you need to think big. The larger your database, the more real estate transactions you'll close, period!

Besides the sales expertise, the size of the database is by far the fine line that separates most real estate professionals from the few highly successful agents in the nation. Top producing agents not only have massive databases but over the years have learned how to leverage each contact and turn them into repeat clients through well-organized and segmented databases.

The "Organize Your Songs" Concept

The organization of your contacts has a direct effect on the success of your business and your life. Accurately profiling or grouping people is a crucial step that most companies challenge to overcome. An organized database will save you time and help you communicate more efficiently with your target market and the people within your network. Whether you want to mail a greeting card to a past customer, email clients, or contact friends across the nation, you must create a profiling strategy to group contacts into distinct categories.

Contacts in your life are very similar to songs in your music library. You can't possibly remember the name of every song you have among the thousands of songs you've downloaded over the years. For instance, if you want to listen to Classics, you'd have to group classic songs into a

Classics playlist. You do so by assigning the genre of each classic song correctly. Similarly, if you want to listen to Pop, Rock, Hip Hop, or Jazz, each will have to be adequately grouped in their specific genre. Songs that don't belong to a playlist with missing genre or the artist name may not play unless you search for them specifically by name.

Likewise, the people whom you don't appropriately group in your database, and the customers who aren't frequently on your mind may go untouched. The better you manage your networks, the higher your chance in succeeding at anything you do in life. After all, most of what we do daily involves communicating with people. Why not organize and streamline communication with others so we can build better relationships and connect with the people who matter most to us?

Organize Your Existing Database

It is less costly to generate revenue from an existing customer base rather than acquiring new ones—especially if you already have a decent size database. However, do not confuse this method with generating leads from sources other than your existing network. Nurturing an existing database to produce more business is indispensable in leveraging already-developed relationships, but this doesn't mean you should only focus on your current sphere. Besides cultivating your existing network and building new connections to deliver a consistent flow of leads is essential for your business.

To leverage your existing database, you must organize it. Super agents regularly close more sales by tapping into their current book of business. They leverage the different types of people they know because of their organized databases. Top producers group people into various segments so they can personalize communications, streamline the follow-up process, and strengthen connections. By organizing their database and cultivating relationships, they boost their production and scalability.

If you want to become one of the top 100 agents in the nation, you can't merely add people to your database and hope most will convert to customers. You must target a specific group of audience with a highly focused niche. Rose, for instance, is a top agent because she focuses on a particular farm and caters to sellers only. It doesn't mean she will turn down other opportunities—and yes, some agents do refer buyer leads out—but she is all about sellers. She dominates the seller's market in her farm. That's her niche.

If Ross wants to become a top real estate agent, he can't be a *"Jack of all trades, master of none."* Generalizing is a major mistake that most real estate professionals make. They want to serve sellers, buyers, For Sale by Owners (FSBOs), and owners of expired listings. Furthermore, they want to receive referrals from the few customers they already have and work with partners with whom they rarely communicate. Agents who generalize due to their fear of losing potential opportunities end up losing far more opportunities than those who specialize in a specific target audience with a particular niche. Specialize by targeting one specific audience or a distinct group of people while building your database—don't generalize.

Besides building a massive database and targeting a specific audience, Ross must tag contacts with their distinct roles. By Tagging people, you organize them into different groups. This process is called profiling. Profiling contacts of each particular audience, and segmenting them into multiple personas will help Ross focus on each specific niche. For example, when Ross separates buyers from sellers and first-time homebuyers (FTHBs) from move-up buyers, he can easily target each persona with personalized real estate services.

To dominate your marketplace, you must also profile your database. Isolate family and friends. Identify your potential buyers from sellers, and then separate FTHBs from luxury buyers, FSBOs from Expireds, and vendors from partners. This way, you can personalize your brand messages to each specific group and invest in sales and marketing efforts that will lead to inevitably more closings.

To correctly tag people, Ross must go through his database and manually categorize each contact with the proper type of audience. I highly recommend you do the same and tag every single contact in your database. Profiling connections and organizing the database is the first step to effective communication and streamlined growth. This task can't be outsourced nor delegated because no one knows your contacts more than you do. Invest the time to tag your network correctly. You'll be glad you did. All your marketing and sales campaigns will be dependent on it.

> *Take the Next Step*
> *Execute Tactic 7.0: Organize Your Existing Database*

Build a New Database

Marketers, salespersons, and real estate professionals, Ross, and Rose add two types of people to their database: either people they know or with whom they are familiar.

First, they add close people they know, such as their Sphere of Influence (SOI). For example, family, friends, neighbors, past clients, and partners. Close contacts generally account for a few hundred people. While that might be an acceptable sized database, neither you nor Ross will have a sustainable real estate business with a network that comprises of only hundreds of people.

Second, they add people with whom they are familiar. Those could be individuals they've met at network meetings or leads acquired through various marketing efforts. Also, these prospects may account for another few hundred people. Again, you will have a zero chance at dominating your marketplace with only a few hundred people in your database.

While everyone can still pursue these two methods, today, I will introduce to you to an innovative way of building a massive database and a thriving business. Instead of going after the few leads acquired through networking and online marketing, you can go after a substantial number of a specific audience. Instead of pursuing them to do business with you, people will come to you first. Instead of trying to get to know them, they will get to know you first. Building a database at a massive scale is executed by leveraging the people that you don't know and that don't know you.

Three Networks to Scale the Magnitude of Your Business

Let us define the ways to build your database and grow the magnitude—mass—of your business:

1. In-network: Through people that you've already met.
2. Network: Through people that you're familiar with and may not have met yet.
3. Out-of-network: Through people that you're not familiar with and certainly have not met yet.

In-Network: People You've Met

These are the people that you know, and they know you.

- Sphere of influence (SOI)
- People that like you, love you, and trust you.
- Family and friends
- Past clients
- Partners, affiliates, and vendors

- Colleagues
- Acquaintances

Network: People you're Familiar With

These are the people that you may know, but they may not know you.

- Prospects: Cold leads in your database
- Leads: Warm leads from open house events, local events, and other network meetings
- Opportunities: Hot leads in your database
- Deals: Contacts in your pipeline that are in the process of becoming customers
- Random: Who the heck is this? (Yes, we all have those contacts)

Out-Of-Network: People You Haven't Met

These are the people that you have not met, nor have they met you.

- Your farm
- Sellers
- For Sale by Owners ("FSBOs")
- Expired listings ("Expireds")
- Short sales
- Probates
- Distressed property owners
- Foreclosures
- Buyers
- First-time home buyers (FTHBs)
- Relocation buyers
- Move-up buyers
- Luxury buyers
- Everyone else

Most agents rely on their sphere and referrals as their primary source of business. With an In-network size of an average of five hundred people, it will be challenging to receive an immense amount of referrals where these agents can dominate their marketplace. Some will barely make it, and most may fail.

How many referrals do you think Ross will receive monthly from the 200 contacts in his database? Not much, right? If your real estate business model is relying solely on referrals from your limited SOI, you and Ross MUST make a shift immediately.

On the other hand, some agents depend entirely on lead generation targeting people they haven't met. Most agents target Out-Of-Network audiences, such as Zillow. Others pay a high price tag for online ads to attract new clients. Focusing on Out-Of-Network ignoring the other two most popular networks will keep these agents busy, but they will not dominate.

To scale the magnitude of your business and dominate, you must work all three networks. In other words, you must get as many people from Out-Of-Network into the Network and eventually into the In-Network. Thus, the main focus for lead generation should begin with the people that you don't know, not the people you already know.

Big real estate producers capture Out-Of-Network leads by investing in online advertising. They do not knock doors, but they consistently send Every Door Direct Mail (EDDM) campaigns. Besides content marketing and blogging, they are also actively engaging on social media platforms, and so should you.

I have an affordable solution for those of you with small marketing budgets. Today, I will share a growth strategy that is far more effective, less expensive, and yet, yields better results than spending thousands of dollars on marketing.

The strategy of capturing business from the Out-Of-Network is simple and powerful. Though there is more to it, I am going to give you a straightforward example to illustrate its effectiveness. Instead of targeting 500 thousand to 1 million Custom Audiences on Facebook, pick and target only a specific number—let's say the 6,000 contacts within your farm of potential sellers only. Not only will you save money on paid advertisement, but you will target those 6,000 people with highly personalized and useful messages turning them to loyal clients.

Take a moment to think about this strategy. Instead of paying for marketing campaigns to acquire a few leads every month, you proactively add the maximum number of people of potentially qualified leads into your database. Then market to those people only to convert the maximum number of quality leads to customers.

Building your database through the Out-Of-Network is done through Online Real Estate Farming.

Three Types of Farming to Scale the Magnitude of Your Business

There are two types of farming that are currently being used by most top producing agents; geographical and through paid ads. However, I have added a third method to separate you from the crowd and take your business over the top. Most agents know about traditional real estate farming. Also, agents know about farming through paid ads on social media and syndication sites, but have you ever thought of generating online demand through predictive analytics? Online farming using artificial intelligence is one of the reasons why "Dominate Real Estate" was born.

I am sure everyone reading this book is familiar with the word "farm." While farming means growing crops—in real estate, it means regularly advertising to a specific market segment to build awareness and develop a strong brand presence. Figuratively, it is the process of planting seeds of future business, nurturing them with marketing, and then later—you will reap the rewards in real estate sales.

If most agents knock doors and send direct mail to a specific farm, how can you grow momentum in such areas at a much larger scale? The competition is too steep—and there are only so many doors to knock, and a few communities to which you can send mailed letters. Why limit yourself to door-knocking and leaving flyers in neighborhoods? Besides, door-knocking is time-consuming, and direct mail becomes expensive after a while. How can you do so much more at a much faster speed to create a more significant momentum?

You plant more seeds—way more seeds. You get people—way more people—to help you push that flywheel at a much faster speed. Instead of door-knocking and sending direct mail, I will introduce you to a new method of farming.

Geographic Farming

Let's briefly touch on the subject of how the offline marketing method of real estate geographic farming works. The farming methods that real estate agents know is planting seeds through traditional advertising efforts to raise brand awareness, capture leads, gain listings, and connect with potential clients. For example, with conventional farming, you might:

- Plant seeds by writing cards to family and friends, telling them how much you appreciate them in your life. Similarly to clients thanking them for their business and loyalty.
- Use USPS's Every Door Direct Mail (EDDM) to farm local Zip Codes.
- Research networking events on the web and in your local area, then attend them to network with the local homeowners and

your community.

- Send email requests to potential partners such as loan officers, CPAs, financial advisors, appraisers, title reps, and escrow officers with whom you want to co-brand.
- Set up appointments with past clients to catch up, ask for referrals, and prepare for possible new listings.
- Invest in others by networking with more people to plant more seeds for the life you want to lead.

These traditional efforts work to a certain extent. I highly recommend you do all of them. They work—only if you consistently execute them for a minimum of three months.

> *Take the Next Step*
> *Execute Tactic 7.1: Farm Your Network Using Traditional Marketing*

Social Media Farming

Farming through social media is the most commonly used farming method by tech-savvy real estate millennials and top real estate producers. Instead of the traditional mailed letters and offline efforts like in-person networking events, agents have been widely using Facebook ads to build brand awareness and capture online leads.

Studies show that Facebook is where people connect online and build relationships. According to statista.com, and as of January 2019, the Facebook community has grown to 2.23 billion monthly active users (MAU) leading the market in social networks worldwide. Consider Facebook's market size and the fact that Facebook Ads are one of the most affordable internet marketing channels online. You will do your business a disservice if you don't farm your territory using Facebook.

Facebook's Custom Audience feature allows you to target a particular audience based on specific demographics and geolocation. You can serve ads by city, neighborhood, or zip code. You can also customize your advertisements to people with a particular language, interest, or behavior.

Moreover, you can opt to deliver Facebook ads on a wide selection of networks: choose to serve ads in the desktop news feed, the mobile news feed, Messenger, Audience Network, or Instagram account only. Your options are vast but don't spread your marketing budget too thin. Test different ad placements to find out what works best in your area and with your audience. Keep in mind that running ads for a short period and a small budget doesn't always yield the best results. To maximize your return on investment, you need to run your ad campaigns continuously. Also, you must test different ad combinations regularly. Try performance on various elements such as sales copy, videos, images, and calls-to-action. Ad optimization is an ongoing process that you need to keep iterating over time for the best conversion ratios and the most leads.

While offline geographic farming is necessary, social media has been proven to be more robust. It is much easier to track the performance of leads generated online. Most real estate professionals have been shifting to Instagram due to the rising number of users. Emarketer.com research shows that as of January 2019, more than one billion people use Instagram every month, and 34% of users are millennials, also known as Generation Y.

Shift your focus to social media in farming your area, but don't debase the value of direct mail once a month. It is crucial to keep an optimal balance between geographical and social media promotions. If you are targeting millennials, Facebook and Instagram have a massive potential to generate the most home buyer and seller leads. However, if your neighborhood and your audience aren't Generation Y or Generation Z, then offline marketing might still be best.

> *Take the Next Step*
>
> *Execute Tactic 7.2: Farm Your In-Network Using Social Media Marketing*

Predictive Online Farming

While geographic and social media farming work to some extent, they both use the concept of advertising to the mass and without real estate intelligence. Neither CoreFact's Direct Mail nor Facebook's Paid Ads are performing predictive analytics algorithms on their marketing campaigns. Even with Facebook's Custom Audience and Lookalike demographics, Facebook doesn't identify who's likely to sell or buy a new home. Therefore, while most Facebook campaigns seem to work for many agents, it would do wonders when you upload a pre-targeted contact list of potential sellers and buyers. Besides, it would be much cheaper and highly relevant to use a pre-targeted contact list than Facebook's Network or Custom Audience. For example, would you preferably advertise to 1,000 homes that fit a specific demographic with a somewhat relevant message, or the top 100 potential sellers with highly relevant advice? Sure the later will yield better results and more deals at a lower budget and much less hassle.

Now imagine you can do this on Facebook, Instagram, LinkedIn, Twitter, YouTube, Google, and pretty much everywhere online! That is what I call "Predictive Online Farming" or simply, "Online Farming." Your ability to target relevant prospects online with personalized messages at meager investments. It is probably the most fundamental concept in this book. Get a professional or a virtual assistant to help you implement the online farming strategy.

The Six Steps to Building a Massive Database Using Online Farming

From the momentum concept, you've learned that you need mass amounts with fast velocity to create noticeable momentum. The Online Farming strategy will leverage people, time, and technology to accelerate momentum. In other words, you will emphasize on the importance of leveraging people in *mass*. This way, you can speed up the *velocity* of the online farming process to create unstoppable *momentum* in growing your real estate business.

You can apply the same concepts of traditional offline real estate farming in the digital world, but on a much larger scale, with an in-depth focus on data and intelligence. Similar to conventional offline advertising, online farming includes brand awareness, brand activation, customer acquisition, retention, engagement, conversion, and growth.

Cast a wider net and dig deep to build a massive Out-of-Network database. Execute the subsequent six steps and follow through the next chapters.

1. **Pick Your Territory**: Determine and decide which areas you want to dominate.
2. **Data Collection**: Collect 6,000 contacts from a predefined target audience.
3. **Data Enrichment**: Find contact emails, phone numbers, and social media information.
4. **Data Validation**: Make sure the email addresses and phone numbers of all contacts are valid.
5. **Data Normalization**: Divide contacts with several email addresses and phone numbers into multiple rows.
6. **Data Sanitization**: Remove invalid, government, irrelevant data, and data on the Do Not Call (DNC) list.

> **Note**
>
> *I recommend you pick your territory and have your marketing person or a virtual assistant execute the data related steps 2-6.*

In these six steps, you will understand how data can hugely impact your real estate business. You will also learn how:

1. Data is used to find the best areas to farm your real estate business.
2. Data is collected to build massive databases in less than three months that takes pro marketers over three years to develop.
3. Data enrichment can help you find the contact information of any person in the nation.
4. Data validation can improve communication and achieve a better bounce rate in text and email messages.
5. Data normalization can remove duplicates and take advantage of multiple contact information simultaneously.
6. Data integrity can keep you out of trouble. (Government regulations, server attacks, blacklisted domains, and Internet Protocol or IP addresses)

By executing these six steps, you will build a massive database and a highly organized network. Also, accurate data will lead to higher conversion ratios and increase sales.

Find the Best Real Estate Market to Farm

Before planting seeds, farmers find arable farmland suitable for farming. Otherwise, all the plantings, nurturing, and harvesting will lead to disappointing results. Hence why this is the most important and the most critical step in getting the most out of farming. In this step, you will find the best real estate market to farm. Farming a blanket of beach cities and multi-million dollar homes doesn't necessarily mean that is the best move. When projected inventory is low, and there are twenty other competing agents trying to farm the same area, such a decision may lead to a huge disappointment.

Deciding on which area you want to cultivate is usually done by understanding real estate farming. However, you are not picking a territory to farm through direct mail. While you can, you are going to do a different kind of farming—online farming!

Pick Your Territory

While most agents farm their neighborhoods, sometimes some agents feel there is a need to cultivate outside the few hundred homes in the community. Here are some factors to consider in choosing the best market to farm.

1. Population and Job Growth

Look to farm areas where an indication of population growth and job opportunities are present. For example, new developments and new construction areas attract newcomers, whether being new homeowners or move-up buyers. Be the first to enter and farm these markets. Make sure to cultivate the surrounding neighborhoods even if other agents are marketing to them. It is better to be in the game than not play at all.

2. Generational and Economic Diversity

Pay close attention to the different kind of generations you are targeting. Current generations are culture creators who are redefining essential components in our society, especially in terms of an economic framework—labor and employment. Learn the difference between age groups and the critical role they play in the economy. As of the writing of this book, it would make the most sense to put your focus on farming to millennials[i] and Generation Z. Those are your best potential customers. Baby Boomers and Generation X tend to have close relationships with other real estate agents within the family.

3. Family, Living, and Business Friendliness

Explore great neighborhoods with the best characteristics to raise a family based on public schools, crime rates, cost of living, and family-friendly amenities. People are drawn to live in attractive areas. They look for unique houses and neighborhoods with fun lifestyles. They prefer communities with highly rated schools and tree-lined quiet streets where children can play outside. It is also a plus to have easy access to restaurants and shops. Not to discount natural gas availability, electric utility fees, broadband internet speed, adequate clean water

---

i   *Check NAR for the latest real estate statistics. Perform a Google search for "Real Estate in the Digital Age" and "Technology Survey" by The National Association of Realtors®. Learn about your home buyers and sellers demographics and the latest generational trends. The numbers are shocking. 99% of Millennials search online for real estate information, and over 75% of real estate professionals nationwide don't even have a blog or an intuitive IDX website to draw them in.*

and environment, zoning, and physical infrastructure support, among many other characteristics that attract people and potential clients.

4. The Market Cycle

Most agents stop promoting when the market declines and start the farming process when the demand is rising. By doing so, other agents who cultivate consistently would have more advantage over erratic and irregular agents. Don't be a whimsical farmer, so you don't miss the rising wave. That is why it is essential to farm your territory throughout all market cycles, continuously whether rising, hot, booming, declining, or stable.

5. Familiarity

Mark your townships and work local areas which you know very well. Make "Local Expert" as a part of your unique value proposition. Don't be a *"Jack of all trades, Master of none!"* Gaining familiarity with your neighborhoods gives you an advantage over the non-local agents. Your competence and confidence with your territory give you a unique advantage over other real estate professionals who may not be showcasing such remarkable quality. Make sure you know the history behind your neighborhoods and communities, including recreational parks, golf courses, gyms, schools, malls, shopping centers, and office buildings.

6. Turnover Rate

Look to see how long homeowners live in an area before they relocate. The turnover[ii] rate refers to the average number of houses that sell each given year. For instance, if a community has a total of one hundred homes and an average of ten homes sell every year, then the turnover rate for that neighborhood is 10%. In choosing the right area to farm, you need to aim for a turnover rate of at least 7%. Be strategic about your farming method. The higher the turnover, the better your territory.

7. Competition Market Share

Competition market share refers to how much percentage your competition has in the market. For instance, if there are ten sales a year and one agent represents six of those transactions, then that potentially

---

ii   *Turnover Rate is the number of homes sold in a specific area, divided by (÷) the number of total households in that area.*

leaves only four other listings for you to win. While you still have an excellent chance to land a few opportunities, tapping into a highly competitive neighborhood with over 50% market share will cost you bigger marketing budgets and extra sales efforts. You would probably be more fortunate and better off avoiding regions where competitors have a significant market share, especially if only one agent has over 50% of the market.

There are other factors that you may want to consider during your competitive analysis—for example, the average sales price, days on the market, and inventory levels. Average income, average age, and transportation options also matter. Whatever data points you want to consider, run the numbers, and research them. This way, you invest in the right neighborhood that can give you a good return on your money. To check on data updates within this section, visit the websites for the Census, CityData, and MLS.

> **Take the Next Step**
> *Execute Tactic 7.3: Find the Best Real Estate Market to Farm*

Data Collection

Data collection is the very first step in growing your database. Instead of waiting for leads to come to you—which is a costly and a slow process—you will collect potential clients ahead of time.

There are various lead sources that you can use for the different target audience. Nevertheless, you should always use multiple sources and never rely on only one source of leads to power your business. Top producers are consistently on the lookout for the new and trending best lead source in the industry.

The performance and reliability of lead sources will vary depending on your target audience (sellers, buyers, FSBOs, or Expireds) and your geolocation (city, suburbs, industrial, or commercial). You should always have a minimum of two principal sources to compare, which yields higher closing. Keep on adding and comparing more lead sources. The more you test, the more likely you will dominate your marketplace.

Lead Sources

Remarkable real estate producers dominate their markets and stay on top by consistently closing more deals from three to five different sources. They neither rely on Zillow Ads only, nor do they rely entirely on Facebook advertising. Also, while referrals have the highest closing ratio, top agents don't utterly depend on them either. However, instead of counting on one particular origin, they rely on all of them to drive the maximum amount of leads into their pipeline.

The evolution and success of your business can't depend solely on one source of leads—especially if you don't control the source. Putting all your eggs in one basket is a vulnerable business model. You will not be able to withstand market volatility nor protect the consistent growth of your business through one client base only. Always be on the lookout to acquire multiple lead channels to fuel the momentum of your real estate firm. Be equipped to pivot when the real estate market shifts or flexibility becomes a necessity. It would be best if you created success by building reliable lead sources for generating sustainable growth.

There are various methods to attract new leads, collect the information of potential clients and their real estate data. Below are the most common lead sources that top real estate producers use.

Collect Potential Sellers and Buyers Data

Using Predictive Analytics

As of today, there are only a few platforms that use predictive analytics to a certain extent to determine who is likely to sell or buy a home.

Remine

"Remine" is one of the leading platforms in residential real estate when it comes to Predictive Analytics. Remine is an online platform that uses consumer and property data to determine when someone is most likely to buy or sell. While the accuracy of prediction isn't guaranteed, this platform can instantly retrieve past consumer transactions, and score them based on their propensity to sell. They call this attribute the "Sell Score." It has low, medium, and high scores. People with a high score means they are likely to sell in the next 12 months.

You can track properties, neighborhoods, and people so you can stay informed with real-time data. Search for any property, community, school zone, or postal code, and see your results on an interactive map. Once you have determined the market you want to track or farm, you can purchase household contact information to connect with your potential seller leads.

At the writing of this book, your MLS subscription includes the "Starter" plan that gives you free access to "Remine." This Starter plan will allow you to track and download up to one hundred households. Check with your broker, MLS subscription, or Remine for accessing additional features and services.

Besides the opportunity Remine provides to discover potential sellers with the "Sale Score" feature, I am a huge proponent of their "Absentee" filter. By filtering Absentee owners with occupying renters, you are finding your potential homebuyers—specifically your first-time homebuyers. Reach out to these potential buyers. Ask if they are tired of renting and if they are in the market to buy. Also, craft personalized ads, letters, emails, and SMS messages with special programs available for first-time homebuyers only.

Why Remine? Most agents are currently using Remine as a tool to send targeted direct mail and as a streamlined door-knocking route. While you can also do so, you will be using information obtained from Remine to segment your contacts into potential sellers using the Sale Score filter. Similarly, organize potential first-time homebuyers using the Absentee filter. Later, you can send Push Notifications, SMS Messaging, and Email Marketing campaigns to these groups. Targeted online campaigns are by far way more effective than the traditional direct mail and door-knocking.

Tip

I highly recommend you have a Remine Pro plan so you can receive households contact information such as their emails and phone numbers. Even if you get personal information of only 50-60% of the overall contacts, a professional plan will save time and money. At the time of this writing, a Pro plan is $199 per month. You may want to consider the option of co-branding with your lender partner to share some of the information from Remine, and split the cost of a Pro plan shall you decide to upgrade.

ProspectNow

"ProspectNow" is another platform that provides both residential and commercial real estate data similar to Remine. They are both very comparable in services and usability.

ProspectNow has a predictive algorithm that helps you identify properties that are most likely to be listed for sale in the next 12 months. Their algorithm looks at hundreds of data sets each week of sold properties. Then they apply those characteristics to properties that have yet to sell. This way, they can determine the score of which homes are likely to be listed next.

Try ProspectNow for three days free. Compare with Remine and decide which one will best fit your needs. In the end, you need to export 6,000 contacts are likely to sell or buy from either platform.

Audience.ai

Another source for collecting a precise target audience using predictive analytics and artificial intelligence is Audience.ai. I am confident that soon, there will be many more companies that offer the same services online. Do your research and choose the source and the company that works best for you and your budget.

Disclosure

I am in no way affiliated with any of the companies or technologies mentioned in this book.

Using Web and Real Estate Data Providers

There are other ways to obtain consumer and property information to farm potential sellers and buyers. Please check with your broker for the best method to access public information as they may vary from one state to another. Most states might have direct access to third-party providers where you may not need to pay for additional services. For example, through your Multiple Listing Service (MLS) or Commercial Information Exchange (CIE) subscription. Here are a few trustworthy services to choose from in collecting data or creating client's lists:

- **Propertyradar**: For $79 per month you can create unlimited lists and export up to 10K contacts.
- **CRS Data**: For $45 per month you can export up to 2500 contacts within one county.
- **ListSource by CoreLogic**: A reliable property intelligence service. You can access data from renters to homeowners, commercials, and everything in between.
- **DataTree by First American Title**: A dependable real estate data solution. DataTree is similar to CoreLogic. They were one company at one point. They provide the same services.
- **Experian**: Build your potential sellers or buyers list based on accurate financial data.
- **Online Data Providers**: InfoUSA or Hoovers by Dun &

Bradstreet.

- **Public Records**: Consider other real estate or public record data providers in your state such as title companies, or city halls.

Collect Foreclosures, FSBOs and Expired Listings Data

There are a few more resources that you can use to collect real estate consumer data. They all provide data on Vacancy, Flippers, Tax-Distressed, Pre-Foreclosures (such as a Notice of Default, a Notice of Trustee's Sale, and a Cancellation of the Notice of Sale), as well as commercial real estate data, Probate, For Sale by Owner (FSBO), and Expired Listings.

You can research each of the following providers and decide which platform works for you: Black Knight, Realist by CoreLogic, MojoSells, SalesDialers, TheRedx, Vulcan7, ArchAgent, LandVoice, ColeRealtyResource.

Collect Commercial Real Estate Data

While the commercial real estate industry is somewhat different from residential real estate, the concepts for lead generation and acquisition remain the same. To convert any potential real estate lead to a loyal client, you will need to have access to reliable data to make insightful decisions. Use the following commercial real estate resources to create your list of contacts and build your massive commercial real estate database.

- LoopNet: Curate databases of commercial real estate leads
- Reonomy: Collect commercial real estate leads
- CoStar: A complete source for commercial real estate intelligence, insights, data, and tools
- Retail Lease Trac: Access to databases of thousands of currently expanding retailers with direct contact and expansion plan info

Do not purchase data from unknown companies or harvest such information from the internet. Consumer and real estate property data obtained from unreliable sources tend to be outdated and, in most cases, inaccurate.

> *Take the Next Step*
>
> *Execute Tactic 7.4: Data Collection: Collect 6,000 Contacts from the Best Real Estate Market*

What Not to Collect

Downloading contacts from social media websites is a standard method most marketers and real estate agents use to increase the size of their database. Your Facebook fans and friends are followers, not subscribers. Your LinkedIn connections are business professionals, not leads. Sending mass messages to your social media connections that didn't opt-in will hurt your business. While you may occasionally get a callback, you will upset far more people with your email newsletters.

There is a better strategy to connect with your followers properly. It is called a Direct Outreach Strategy where you personally connect with each person to build brand awareness and get authorization to contact them.

While I don't mind, some real estate agents send me newsletters to which I didn't subscribe. They either follow me on Instagram, are fans on Facebook, or connections on my LinkedIn. Those agents merely added my email address to their CRM, scheduled some automated email campaigns, and off they go. These agents never requested an authorization to email me, and I certainly have no idea who they are. Even worse, and not intentionally, most of their emails land in my SPAM folder. Requesting to connect with me on LinkedIn so that they can put me on their newsletter campaign isn't a good practice. They are not going to earn my business with such a strategy. Instead, doing so will annoy most people and hurt the agent's brand image, domain rank, and the reputation of their IP addresses. That is why the majority of those agent's emails go to the SPAM folder.

Do not SPAM your social media followers. Do not add your LinkedIn Connections to your CRM and start emailing people. Your lead source must be a target audience—potential buyers or sellers, not every fan on your Facebook page. Also, you are not going to put them on a newsletter and hope they call. You will be wasting your time and energy. You don't want to be like most agents in the market. It's sad, but that is what most of them do. Do yourself a huge favor and do not do this. If you insist, at least call your network, introduce yourself, and request permission to message them.

Prepare Your Farm for Growth

Now that you've found arable farmland, you'll need to plow it before sowing seeds. Preparing your farmland for growth by turning your land before planting the seeds is a critical step that gets skipped in most businesses.

What does "Plow Your Farmland" mean? It means, "Optimize Your Database for Conversion." In other words, prepare your contacts to become clients. To cultivate meaningful relationships and convert as many people as possible to raving fans, you need to make sure all contacts in your database have valid information.

How many times have you received an online lead only to find out it was invalid after your email has bounced? Have you tried texting a client only to find out it was a landline? Ran an email campaign and saw a 10% bounce rate? This list of errors goes on! Optimizing your database for business development, marketing, and sales campaigns is the first step to running a smooth and successful business. Many businesses skip over this highly critical requirement and pay a high price later in the sales process just because of the wrong contact information.

The process of filling in the gaps and completing the missing personal information is called "Data Enrichment." The method of cleaning the integrity of the data and making sure it is up to date is called "Contact Hygiene" or "Data Sanitization."

How do you enrich your data and find people's email addresses and mobile phone numbers? Let's address this with a couple of data enrichment options.

Data Enrichment

Data enrichment is the process of merging third-party data from an external authentic source with an existing database. By adding reliable data to your contacts, you gain more in-depth insight into your clients' lives saving on marketing budgets and sales efforts. With access to more data, you can leverage personalized messaging and have better conversions. Hence, data enrichment is vital to closing more sales, and to the success of your real estate business.

There are two types of data enrichment, demographic and geographic. The demographic data enrichment process involves acquiring new data, such as ethnicity, age, gender, marital status, educational and income levels. The geographic data enrichment method consists of land use, zoning, traffic, and adding physical address information.

However, the types of demographic and geographic data are vast. You could receive datasets that include types of houses owned, number of occupying children, their ages, vehicles driven, city boundaries, and mapping insights. While such information may be helpful during the qualification process, obtaining personal data is vital in improving online marketing strategies. Getting email addresses, phone numbers, and the mailing addresses will benefit in sending various marketing campaigns.

Moreover, data enrichment isn't a one-time process that you perform once and then never do again. Customer data is continually changing! For example, people's income levels may rise or fall when they change careers. Their names or physical address may differ when there is a change in one's marital status. Given the likelihood of all these reforms, data enrichment processes need to run continuously. Otherwise, you will risk having outdated information that could lead to customers receiving extraneous marketing and sales offers because your data is obsolete. Keeping personal information current is an enormous undertaking, especially for massive databases. According to a 2016 Study by TDWI, more than fifty percent of companies spend more time cleaning their data than actually using it.

To keep data recent, one must automate the data enrichment process. Machine Learning (ML) algorithms that run continuously can substantially streamline the enrichment operation. Automated enrichment using ML leads to high accuracy in data because computer-operated methods can match and merge data much faster than humans. Ultimately, automating such behavior allows businesses to sustain a high level of enrichment that leads to enhanced customer engagements.

One-Time Contacts Enrichment

Use an internal marketing employee or a virtual assistant to enrich data for your database. Delegate all data related tasks or hire a professional. Don't pay an hourly rate! Outsource the entire project at once or get a professional quote to enrich all 6,000 contacts at once.

Hire a Virtual Assistant or a Freelancer (Recommended)

Use one of your favorite freelancing websites to enrich the 6,000 contacts with mobile phone numbers, email addresses, and social media profiles. Also, improve your existing database if it is missing personal information.

You can choose one of the following services to hire a Virtual Assistant (VA): Freelancer, Zirtual, My Out Desk, or Hire My Mom.

My preferred platform is Freelancer as it has the best quality of freelancers at the perfect price with choices. You have the option to either post a project or start a bidding contest. You can always choose to either pay a fixed price (advised) or pay by the hour (NOT recommended). Best of all, for around $9.50 Freelancer can have a recruiter find you the best freelancers for your project, so you don't have to. It can't get any easier than that!

I would have recommended other popular platforms such as Upwork and Fiverr. Upwork is the most expensive and Fiverr is the cheapest. Besides, the quality of freelancers and their work wasn't the best I've personally experienced with these two platforms. Nevertheless, pick whichever platform you feel comfortable with and stick to it.

Sign up for Freelancer.com
1. Click on "My Projects" next to the logo and then "Create One."
2. Fill in project details and click "Next."
3. Choose for required skills and click "Next."
4. Choose for "Post a Project" or "Start a Contest."
5. Choose "Pay Fixed Price."
6. Select a dollar amount for your budget.
7. Select for a "Standard Project" or "Recruiter project" or any of the other advanced options such as NDA, Featured, Urgent or Private project.

Sign up for Upwork.com
1. Hire for a Project.
2. Job Category.
3. Admin Support.
4. Personal / Virtual Assistant.
5. Add your job descriptions and details (Database Enrichment: Find mobile phone numbers, email addresses, and social media contact information for contacts in my database).
6. Follow the rest of the instructions to post your project.

Once you evaluate and decide on an outsourcing platform, hire six virtual assistants (VAs) each to enrich 1,000 contacts. Each VA should find emails and phone numbers of about 80-100 people per day. You should be able to have all 6,000 contacts completed within a maximum of two weeks. Once you obtain your 6,000 contacts, you'll need to upload them to Google Sheets and share with your virtual assistant. Split your 6,000 contacts into six sheets, 1,000 each.

Please do not underestimate this step! While you might think hiring only one virtual assistant is sufficient, the less virtual assistants you have, the longer this entire process will take. Just remember the Momentum equation; more Mass (more people) and higher Velocity (faster speed) will generate more thrust. The more significant the momentum, the quicker you will acquire and close more clients.

Free Contact Enrichment Resources

Here are some free resources for your virtual assistants to use in enriching your 6,000 contacts: Fast People Search, Find People Search, True People Search, Carrier Lookup, Free Carrier Lookup, Fone Finder, Text Magic, Data 24-7, and ZLookup.

There are no guarantees that these free resources will always be available. Also, there might be a limitation to their use. Hence, I will share with you paid enrichment services that you may want to consider should the free resources deem insufficient.

On-Going Data Enrichment

You can hire a professional company to enrich contacts in bulk and all at once. Shop around for the most affordable rate. Most of these companies use the same sources of data and tend to have a 50% to 60% match enrichment rate.

Professional Contact Enrichment Services

There are other data enrichment methods to choose from than hiring freelancers. However, they will cost more money. The more data you need to append, the more they will cost. I want to make sure you are aware of these options should you need them: DataFinder, Revaluate, Speedeon Data, Intellias, Intelius, Geographic Farm, Melissa Data, Datanyze, ClearBit, Pipl, FullContact, and LexisNexis.

To append consumer data such as landlines and mobile phone numbers to your prospecting lists, use one of the following services: Skip Genie, Batch Skip Trace, or Need To Skip.

> *Take the Next Step*
>
> *Execute Tactic 7.5: Data Enrichment: Enrich Your Entire Database of the 6,000 Contacts*

Data Validation

Data validation is a process of validating data types within your database; not to confuse the process of validation with verification. We will discuss verification in the last step under "Data Sanitization." However, validation, for example, is merely assuring all phone numbers are in the format of "123-456-7890" or "123.456.7890" or "1234567890" or "(123) 456-7890". From reading these numbers, we do not know if they are verified—meaning, will 123.456.7890 ring when you call it? However, its format is valid as a US phone number; thus, this phone number is valid regardless of whether it is available or verified. Similarly, email@gmail. com is a valid email format, but most likely not reachable (confirmed to exist, or fake).

Make sure your VA checks the following data attributes to ensure they are valid for all 6,000 contacts within your database and your existing database.

1. Property owner first and last names are not empty.
2. Property address information is not empty.
3. Email addresses do not have spaces or invalid characters. Also, emails must not be empty.
4. Phone numbers are digits such as 1231231234. (No dashes, spaces or parenthesis). The phone number field is not optional. It cannot be empty because Facebook and Instagram ads rely heavily on mobile phone numbers within your database.

You can validate these fields using a spreadsheet through the "Filter" feature. Again, hire a virtual assistant or have a professional clean up all contacts from invalid data. This process will save you money in the Data Sanitization process, marketing campaigns, sales campaigns, and in the many years to come.

> *Take the Next Step*
>
> *Execute Tactic 7.6: Data Validation: Validate Personal Data of the 6,000 Contacts*

Data Normalization

Data Normalization is a complex organizational process a data scientist undergoes to sort a database in a functional format. Jumping into what matters, leaving the technical part out of this section, there are two reasons why you are learning about data normalization:

1. To take advantage of every email address and phone number you have in your database
2. To remove duplicate contacts and duplicate information within your contacts

Through the many years of experience working with lenders, brokers, agents, and many other large companies, I have come to realize that most deal with major database issues. Not only do these issues cause an unrealistic shift in reports and metrics, but they also lead to frustrations and lack of productivity. Dealing with an unhappy customer who never received your email, or has received your message four different times is not pleasant—and indeed not professional.

Have you come across a situation where you want to email a client that happens to have a couple of email addresses? Which email address does Frank prefer? What do you do? Let me guess, send the email to both email addresses in your CRM. How about when you want to call Sally the client and then there she is saved twice on your phone? Which record has the right phone number? What if you're going to send an email campaign or run a Facebook Ad with Custom Audience? Which email address do you use for Frank, and which phone number do you use for Sally? These are common issues most people face! Contact hosting platforms such as iCloud or Google Contacts save our personal data in what is called a "Denormalization" format. This format is where multiple data sets belong to a single record—for example, two different email addresses belonging to one person.

When it comes to your CRM or any other form of communication, it is a good rule of thumb to use the personal email address as the primary email—unless the customer indicates otherwise. When it comes to marketing, you want to take advantage of all available email addresses and phone numbers. However, you must remove duplicates so your messages—emails and text—aren't sent multiple times to your clients.

To solve the disadvantage of multiple data and remove duplicate information, you—or someone you hire—will need to convert *denormalized* contact information to *normalized* contact information.

Here's a practical example that outlines all of this—what might seem like nonsense will soon start making sense:

Denormalized Table

ID	Name	Personal/Primary Email	Work Email
1	Frank Ferrell	Frank@personal.com	Frank@business.com
2	Sally Smith	Sally@personal.com	
3	Sally Smith	Sally@personal.com	Sally@business.com

Table 1: Notice duplicate records for Sally@personal.com. Most CRMs and Email Marketing Platforms will send emails to the Primary Email, which may result in duplicate messages to Sally@personal.com and not

reach her business email.

Normalized Table

ID	Name	Email Address	Email Type
1	Frank Ferrell	Frank@personal.com	Personal
1	Frank Ferrell	Frank@business.com	Business
2	Sally Smith	Sally@personal.com	Personal
2	Sally Smith	Sally@business.com	Business

Table 2: With a normalized table, all emails are unique with no duplicate records. Also, you will deliver messages to all email addresses in your database.

Follow the same normalization process for other columns such as multiple phone numbers or several mailing addresses. Add another column for "Primary" to denote which email address, phone number, or physical address is the client's preferred record (See *Table 3*). Also, when your contacts have unique email addresses and mobile phone numbers, you will have rewarding results in your marketing and sales campaigns. Besides, text messaging and email communication will be more productive, and your collaboration with clients will be seamless.

Table 3: Denoting primary email and phone records

ID	Name	Email Address	Phone Number	Primary
1	Frank Ferrell	Frank@personal.com	(111) 111-1111	Yes
1	Frank Ferrell	Frank@business.com	(222) 222-2222	No
2	Sally Smith	Sally@personal.com	(333) 333-3333	Yes
2	Sally Smith	Sally@business.com	(444) 444-4444	No

Take the Next Step

Execute Tactic 7.7: Data Normalization: Normalize Your Database Information of the 6,000 Contacts

Data Sanitization

Data Sanitization is also known as *Contacts Hygiene* (the integrity of data). Contact List Hygiene is the practice of cleaning an email list to ensure all email addresses are active, engaged, and deliverable. Regularly implementing and maintaining this manner removes unengaged contacts—including their email addresses—from your marketing list. This act will decrease risks for SPAM traps, complaints, and invalid hard bounces. It will also improve domain reputation and email deliverability.

Perform the contact list hygiene process periodically to improve email deliverability, remove SPAM traps, and successfully communicate with inactive contacts.

Now, take the comma-separated values "CSV" file with the 6,000 contacts and your existing database, then clean up the property addresses, email addresses, and mobile phone numbers using **Cylinder.AI** (*http://cylinder.ai*). Alternatively, the following services:

- *Clean Everything: Cylinder.AI or APILayer*
- *Mailing Address Verification: Google Earth*
- *Email Verification: Zero Bounce, Never Bounce, and Data Validation*
- *Phone Verification: XVerify, NumVerify, Data Finder, Text Magic, Phone Validation, and Twilio.*

Disclosure: The Do Not Call (DNC) List

While the DNC law may or may not apply to real estate agents as they are not telemarketers selling goods, I highly recommend checking your records against the DNC list before calling consumers. The fact that most consumers will think you are a telemarketer, it would be in your best interest to save yourself the hassle and not contact those people, period.

After your VA verifies that your 6,000 contacts have mobile phone numbers, make sure to remove phone numbers registered to the Do Not Call list. See the Federal Trade Commission's website, DoNotCall.gov website, Do Not Call Compliance, or RealPhoneValidation.

How much does it cost to access the DNC registry?

According to the Federal Trade Commission website, data is free for up to five area codes. After five, the annual fee is $63 per area code of data up to a maximum yearly price of $17,406.[iii]

Take the Next Step

Execute Tactic 7.8: Data Sanitization: Perform Contacts Hygiene Process to Sanitize Your Database

---

iii   Source: ftc.gov; Federal Trade Commission website.

"The number of people within your network is the momentum of your business and the magnitude of your success!"

—James Tyler

Calculate to Dominate

Hypothetically, after your virtual assistant cleans the 6,000 collected records in your database, suppose only 3,600 contacts had valid email addresses and phone numbers. The question remains if you did marketing and sales correctly, will the 3,600 contacts in your database generate the annual income you're seeking?

On average, the annual conversion rate for marketing and sales is 2% (worst case scenario).

Thus, 3,600 x 2% = 72 potential transactions per year.

Then, 72 / 12 = 6 potential transactions per month.

If the number of potential real estate transactions of six per month isn't enough, collect more contacts as described in the "Data Collection" step. Go through the data enrichment process and the contact hygiene process until the anticipated number of potential transactions closed per month makes sense before moving on.

If you need to know how many more records you need to collect, it will help to identify your validity ratio. For example, divide the number of valid contacts (3,600) by the total number of people in your database (6,000). You will then determine the validity ratio. This way, you know if you need to double your transactions, you'll need at least another 3,600 valid contacts or 6,000 more household records. For instance, to get your database to a total of 6,000 clean contacts, you will need to collect an additional 4,000 besides the 6,000 you've already gathered.

1. Initial Collection: 6,000 household records.
2. Validity Ratio: 3,600 validated / 6,000 total = 60% validity ratio.
3. To obtain a total of 6,000 valid records: (6,000 - 3,600) / 60% = 4,000 needed household records.
4. 4,000 x 60% = 2,400 valid contacts.
5. 3,600 + 2,400 = 6,000 total valid and clean contacts in your database.

This example is a hypothetical scenario to demonstrate validity ratio and the process necessary to work your numbers to fit your desired income. Now, let's determine potential transactions and commission at a 6,000 sized database of clean contacts and a 2% conversion ratio, which is the industry average.

1. 6,000 x 2% = 120 potential leads per 90 days. (Assuming no further growth all year round for the sake of the example). In a real situation, you should be growing the database every month, which should result in much higher sales and bigger commission checks.
2. 120 / 3 = 40 potential leads per month for 3 months
3. Suppose you only close 20% of the leads: 40 x 20% = 8 transactions per month
4. Eight transactions per month at an average sales price of $350,000 = $2.8M
5. $2.8M x 2.5% commission = $70,000 monthly income (The Absolute Minimum).

$70,000 revenue per month is the magnitude you will create from the momentum of having 6,000 clean household records in your database at 2% conversion ratio and 20% pipeline closing ratio. Such income should be a short-term goal to achieve within 90 days from reading this book.

Do the math, know your numbers, plan, strategize, and execute. Do you see why I have mentioned 6,000 contacts over 35 times in this chapter? So you can close a minimum of eight real estate transactions within the first 90 days and then scale every month exponentially.

"Calculate the momentum of your business to dominate your marketplace."
—James Tyler

Ask yourself, "What is the current momentum of your business? What is the number of potential real estate transactions you anticipate closing monthly?" In other words, what is the current size of your database? Moreover, what kind of magnitude are you generating? The scale of your business's momentum towards success is as large as your database. I just showed you how to grow your database up to 6,000 contacts within a maximum of two weeks period. You can follow the same process to build your database from 6,000 to 12,000, 24,000, and 48,000 to 100,000 contacts. The larger your database, the more significant the momentum, and the more transactions you'll win. The more clients you will get, the more referrals you'll eventually receive—which will grow your revenue even more. Domination is a process that starts with building a massive database. Get on it.

Regardless of the database size of other agents—whether it is 500 or 2,000 contacts, you need to set a minimum initial target of 6,000 clean records to dominate your territory. Once attained, make your 6,000 contacts the baseline number and scale from there. Strategically planning is a must-do step if you are serious about succeeding and dominating your real estate marketplace.

If you follow this step alone, you will be guaranteed to close a minimum of ten transactions per month in the first 90 days executing all concepts shared in this book. Based on your efforts, you will exponentially increase closed sales and grow far more revenues.

> **Take the Next Step**
>
> *Execute Tactic 7.9: Calculate the Momentum of Your Business to Dominate Your Marketplace*

CHAPTER EIGHT

Plant More Seeds

"Without continual growth and progress, such words as improvement, achievement, and success have no meaning."
—Benjamin Franklin

Rose closes fifteen to twenty real estate sales per month. Ross does one or two but desires to be like his wife and close more transactions. To add more listings to his pipeline, Ross signed up for Facebook and Zillow ads. Unfortunately, he did not compare to what she was closing. Thus, he moved his efforts to email marketing.

Ross added all of his contacts from his LinkedIn and Facebook accounts to his newsletter. He emailed people for a few months in hopes for a listing. To his disappointment, he discovered that people weren't opening his emails. Some users unsubscribed, and others marked his messages as SPAM. Ross hit a brick wall with no deals! Where did Ross go wrong?

Most of Ross's connections didn't know him. They weren't expecting his emails because they never subscribed to receive his newsletter. Also, those who are familiar with him, either have not heard from him in a while or found his material irrelevant.

The issue Ross faced is common! Most agents seek to gain leads before generating demand. Demand Generation is nonexistent among small businesses, especially real estate professionals. Therefore, he was experiencing difficulties engaging with people, especially with whom he did not possess a relationship. Just because people connect with you on social media, it neither means that they opted-in to receive your emails, nor will hire you as their agent.

To solve Ross's problem, he must build awareness by connecting with people and joining different communities. He has to provoke demand by presenting relevant values. Ross will convert individuals to clients once he earns their trust. For example, extending a real estate selling guide to Ron the Renter would be insignificant as Ron doesn't own a home. Alternatively, connecting with Ron to understand his desires before the value offering is paramount. For instance, Ross will resonate with Ron by offering him a first time home buyer guide. Showing available low down payment programs and following up with him, Ross will eventually turn Ron into a happy homeowner.

The business development department is responsible for using demand generation to figure out potential clients' motivations and intentions. The marketing department is responsible for driving interest in a company's products or services. Without demand, it is unlikely to produce noticeable awareness, and without engagement, it is difficult to create sales. Thus, revenue begins with demand generation.

This chapter shows you how to establish awareness through demand generation. A useful method to follow is AIDA (attract attention, maintain interest, create desire, and get action), developed by Elias St. Elmo Lewis.

You can use the AIDA model as the backbone for structuring an organization's marketing and sales funnel. The funnel helps sales personnel to further target a potential client by basing their actions on the client's position in the process. For example, your activities as an agent are different with leads ready to buy a home next month versus those who are thinking of buying next year.

Salespeople should seek different sales objectives for their suspects, their prospects, and their clients. The aggregated information from the sales funnel allows a firm to construct an insightful overview across sales representatives and departments to deliver better-structured sales forecasts. Thus, to build a solid sales foundation, the first step is to attract people's attention by creating awareness and introducing yourself. For example, by planting a seed in every potential client that is looking to buy or sell in the next 6-12 months!

The Plant More Seeds Concept

Although planting seeds seem like a daunting task, begin by planting one seed at a time. Continue this step, until you feel satisfied with the amount of seeds you plant. Nurture your seeds by watering them, until a whole garden of green leafy dollars greets you one day. Nothing is more exciting than seeing your seeds blossom.

Here are a few ideas to keep in mind as you plant seeds and farm your own business:

1. Plant More Seeds

Seeds take time to burgeon and flourish. You can't grow a whole garden overnight, and neither will a company. The more seeds you plant, the larger your garden will be. Continually focus on the positives, and soon, you will see little seeds peek out everywhere into something new and promising!

2. Nurture Seeds to Grow

A seed can't stay a seed forever. When cared for with healthy soil, rays of sunlight, and water, they shape into what they were meant to be.

3. When Seeds Flower, Everyone Recognizes Their Beauty

After a seed transforms, everyone appreciates it for what it offers. Each seed has something new to offer and contribute. How did it get to this point? The answer lies with time and transformative change.

Go out there and plant seeds in your business, no matter how difficult it may seem at the moment. With time, patience, and effort, beauty and growth will occupy what was now a pile of empty land.

Just like a beautiful flower attracts butterflies, your business and your life will be a magnet to the right people that will truly make it one worth living.

The first step to turn dirt of land into a beautiful garden is by planting seeds. Similarly, to convert strangers—your 6,000 contacts—to loyal customers, you must directly reach out to them planting seeds through all available communication channels online.

Let's sync your database with online platforms so you can reach out to your contacts and turn your business into a blooming garden filled with revenue.

Sync your Database

Now that all your contacts are optimized, it's time to sync them across multiple devices. Synching your database across platforms can help you convert the highest number of people to loyal clients.

Sync Your Contacts with your Phone

The entire real estate market is moving to mobile. It is imperative that you mobilize your database and your business. To efficiently close more sales, you need to have mobile access to all data, at all times, and on all devices, especially your mobile device. Begin by syncing your contacts in your database with your mobile phone.

Note: Some agents may have an issue adding 6,000 business contacts to their mobile phone. If that is the case, I highly recommend that you purchase a new telephone to use for business.

Add Your Contacts to Google Contacts and Sync Across All Devices

You might wonder, James, why use *Google Contacts* when I already have my contacts in a CRM? What if I do not have a Gmail account? Why should I create a Google account?

Before I discuss the importance of having a Contact Relationship Management System (CRM), please allow me to stress how critical it is for you to have Google Contacts. For those of you reluctant to go through this step, here's why you are doing this:

1. Most of Google's platforms are free to use.
2. Global enterprises trust and use Google services. Google is one of the top secure places to host all of your digital business assets, such as your documents, sheets, photos, and videos.
3. Google is a one-stop-shop that can service all of your needs. Email, Contacts, Calendar, Drive, Voice, Analytics, Ads, Maps, MyBusiness, Webmaster, Postmaster, YouTube, and many more. Visit Google products website to learn more about all that is available.
4. All Google platforms sync together seamlessly and effortlessly, without you having to install any software or implement an Application Programming Interface (API).
5. All Google platforms are in the cloud. They are accessible from any device, anywhere on earth.
6. Your contacts are your database. Having the flexibility of accessing your database from anywhere and on any device is paramount for your business. Google Contacts will provide

you with this convenience, but not every CRM.

All of Google's applications connect back to a Google Mail (Gmail) account, or Google Inbox. Once you add your Google email to your phone; data from all Google cloud applications will pour into your phone. This process will sync your email, contacts, calendar, and Notes. The most critical platform is Google Contacts.

Below are some benefits you will get by regularly syncing your database with Google Contacts:

- Your contacts will have the correct and most-up-to-date contact information across all devices.
- If you ever want to change real estate companies, your contacts and clients' information go with you, regardless of which broker you work with or which CRM you use.
- The most significant benefit for syncing with Google Contacts is so you can easily connect with them on social media. When you are browsing social media applications on your phone, do you ever notice how they suggest connecting with your phone contacts? That's right! Whenever you are online, your entire 6,000 contacts on your phone, and in your database, are right there, on social media.

Tip

Alternatively, if you are an Apple fan, you can use iCloud to sync data across all Apple devices, including your contacts. The same principles apply!

For those who would like to keep a personal phone separate from a business phone, get a new phone so you can use it for business only. I promise it will be worthwhile.

Take the Next Step
Execute Tactic 8.0: Add Your Contacts to Google Contacts

Add Your Contacts to ContactsPlus Mobile App to Enrich Social Media Info

Social media platforms have become a vital communication channel. Nowadays, most people spend a considerable amount of time on social media. It would be in your best interest to stay connected with your potential customers on social media platforms.

While *ContactsPlus*—formerly known as *FullContact*—offers a Bulk Data Enrichment service; they also have a data enrichment paid mobile app. Use it regularly to enrich contact information. I love it and have been using it for several years. At the time of writing this book, the app costs $99 annually. It is probably the best $99 I invest in my business. I have yet to find anything like it. Let's go through the why:

After I enrich my contacts using *Freelancer.com,* perform the *Contact Hygiene* step mentioned in the previous chapter, and add my contacts to Google. I still need to update my contacts' personal, business, and social media information regularly. People switch companies, move around, change, or update their social media profiles. The ContactsPlus mobile app automatically handles all of that for me. It has an essential benefit in the ongoing enrichment process.

Here are some powerful and indispensable benefits you will get from having a ContactsPlus Mobile Application:

- Keep contacts up-to-date and in-sync across all your devices, web, iOS, Android, Gmail, and iCloud.
- Gain easy access to all contacts on your mobile phone.
- Have a safe place to back up your contacts where they will always be available in the cloud.
- Receive added contact insights such as publicly available data like social media accounts, jobs, and photos regularly.
- Merge and remove all duplicate contacts for a cleaner and optimized address book.
- ContactsPlus supports syncing with a wide variety of CRMs.
- Have the ability to export contacts into CSV format that allows you to import them into any external software.

To learn more about Contacts+ and continually enrich your contacts, visit ContactsPlus website.

> **Take the Next Step**
>
> *Execute Tactic 8.1: Sync Your Contacts with Your Mobile Phone and Across All Devices*

Get Instant Access to Public Record Information

Leverage people's information available through readily accessible government websites. There are also professional platforms that can provide you with viable consumer data. Gain access to email addresses, phone numbers, mailing addresses, and property tax records. Use the information available through public record applications to pre-qualify your prospects.

Download one of the paid lookup mobile applications to your phone, so you continually access people's publicly available information on the go. A few great examples of public record providers are Intelius, Been Verified, and Truth Finder.

> *Take the Next Step*
> *Execute Tactic 8.2: Get Instant Access to Public Record Information*

Sync Your Contacts with Facebook

Syncing your contacts with Facebook platforms is one of the most critical steps outlined in this book. By doing so, you will target and advertise to them with paid ads on Facebook Platform, Facebook Ad Network, Facebook Messenger, and Instagram. Social media is where over 80% of people hang out. Do not take this step lightly. It is powerful!

Sync Your Contacts Using Facebook Desktop

Sync your contacts using a web browser on your desktop:

1. Log in to Facebook.
2. Click on "Find Friends."
3. Choose your personal contacts provider. For example, Google Contacts, and follow the instructions on the screen to upload your Google Contacts.
4. Alternatively, choose iCloud Contacts and login to iCloud to sync your contacts with Facebook.

All of your contacts should now sync with your Facebook account. Facebook will suggest newly added people that aren't friends with you so you can connect with them. You will see these suggestions on Facebook's Notification box as "People You May Know." Although you might not know them yet, you will.

Sync Your Contacts Using Facebook Mobile Application

Sync your phone contacts using your mobile device. It would be best if you had Google contacts or iCloud Contacts synced with the 6,000 contacts from your database:

1. Launch the Facebook Mobile App.

2. Tap "Find Friends."
3. Click "Allow" for allowing Facebook to "Access Your Contacts."
4. From the "Setting and Privacy" menu, click "Settings," under "Media and Contacts" section; make sure that "Upload Contacts" is "Enabled."

All of your contacts should now sync with your Facebook mobile app. Facebook will suggest that you connect with the 6,000 contacts that aren't friends with you under the "Contacts" tab.

Adding your 6,000 contacts as friends on Facebook is optional. Since these people don't know who you are, you will use Facebook Messenger to connect and introduce yourself. I don't recommend you add them as friends if you haven't created a prior relationship or formal introduction yet.

Sync Your Contacts Using Facebook Messenger

Facebook Messenger is one of the most effective marketing channels that exist online today. Use it, but don't abuse it. Your contacts are a delicate piece of art. You want to treat them with love and respect. Sync your mobile contacts with Facebook Messenger mobile app to introduce yourself and cultivate relationships.

1. From Chats, tap your profile picture in the top left corner.
2. Tap "People".
3. Tap "Upload Contacts" to turn this setting "On."
4. While you are in the Facebook Messenger App, go to the "People" tab, connect your "Instagram Account."
5. Go back to Messenger home, tap "People," and tap on "Add Contacts" in the top right corner.

You will notice in "UPDATES" that your phone has added your contacts to your Messenger account from your Phone Contacts, Instagram, and Facebook. You'll also see that all the people on your telephone are now available through Facebook Messenger. You can message any of your 6,000 individuals directly through Messenger.

Sync Your Contacts Using Instagram

Sync your contacts with Instagram to connect with people within your territory. Instagram is the most trending social application that exists today! Connect your phone contacts with your Instagram account to promote your real estate services to potential clients in your area.

1. Log in to your Instagram account (regardless if personal or business, but preferably business account).
2. Tap the "Settings" icon.
3. Tap "Account" > "Contacts Syncing" to enable Connect Contacts.

You will notice in the "Discover People" section within the main menu that all of your 6,000 contacts are under the "Contacts" tab. Also, to grow your followers and lookalike audience, Instagram suggests people with similar demographics under the "Suggested" tab. Use your Instagram App to invite and connect with all of your Facebook friends in the "Follow and Invite" section. Under "Follow Facebook Friends," follow all the 6,000 contacts within the "Follow Contacts" tab, respectively.

Facebook and Instagram will suggest all contacts that aren't your friends on either of their platforms. Hence, you now have direct access to all 6,000 connections through both Facebook and Instagram platforms.

Sync Your Contacts with LinkedIn

Similar to Facebook platforms, you want to sync your contacts with LinkedIn so you can connect with business professionals. Active LinkedIn members tend to be owners of luxury homes, C-level executives, investors, and influencers. Most of these high profile clients tend to hang out on LinkedIn more than Facebook, Messenger, and Instagram. You sure don't want to lose out on such a profitable market.

Sync Your Contacts Using LinkedIn Desktop

1. Log in to your LinkedIn account through a browser on your desktop.
2. Click your profile picture.
3. Under "Account," click "Settings and Privacy."
4. Under "How LinkedIn uses your data," click "Sync Contacts."
5. Sync your "Google" Contacts.

Sync Your Contacts Using LinkedIn Mobile App

To import your mobile address book:

1. Tap the "My Network" tab in the navigation bar.
2. Tap "Add contacts" in the top right.
3. A pop-up window will appear asking for permission to find people to connect with from your phone. Tap, "Yes."

4. In the next pop-up window asking for permission to access your contacts, tap "OK."
5. From here you'll have the option to send invitations from your mobile address book.

To sync contacts:

1. Tap your profile picture.
2. Tap the "Settings" icon in the top right.
3. Under the Account tab, tap "Sync Contacts."
4. Switch the toggle "On" (for iOS), or check the box next to "Sync Contacts" to "Automatically find people you know" (for Android).
5. Tap "Continue" on the next screen.

In conclusion, all 6,000 contacts should now sync with your mobile phone, Facebook, FB Messenger, Instagram, and LinkedIn.

You can sync your contacts with other social media applications such as Twitter, SnapChat, or any other future trending applications. For the strategies shared within this book, we will focus on Google, YouTube, Facebook, Instagram, and LinkedIn. Reaching your potential clients on these platforms will account for achieving over 95% of web and mobile traffic.

> *Take the Next Step*
>
> *Execute Tactic 8.3: Sync Your Contacts with Your Social Media Profiles*

What Not to Sync Yet

At this moment, you might think, "Well, since I have 6,000 clean contacts, might as well sync them with my CRM and email marketing platform and market to them." Not just yet! That is what a naive marketer or an amateur entrepreneur would do. While it might tempt to email everyone in your database, please do not do so. Here is why:

You have collected, validated, and verified all emails and phone numbers of all contacts, but these contacts neither subscribed to receive newsletter emails nor SMS messages from you. Before moving these contacts into marketing for lead generation, you will need to create demand generation through business development. It would help if you had manually and directly reached out to each contact in your territory.

Build a direct outreach strategy before syncing contacts with your CRM, marketing automation platform, email marketing platform, text messaging platform, or any other marketing and sales operating-technology you use with your current real estate business.

> *Take the Next Step*
> *Execute Tactic 8.4: Build a Successful Direct Outreach Strategy*

Demand Generation: Direct Outreach

I have witnessed several small-to-medium businesses (SMB) lack the essential department responsible for building the company—the "Business Development" (BD).

One of BD's primary responsibilities is to attract the attention of potential clients by building awareness and creating interest. That is by introducing values and benefits of a company's products and services. For you, as a real estate agent, acquaint who you are and present your value proposition to your target audience. If there's demand, you can then move those people who showed an interest in your real estate services to the marketing department for lead generation, acquisition, and cultivation. Converting leads to customers is a multi-step process. You must plan and execute each step properly within each department from business development, marketing, and sales.

Having a massive spreadsheet of 6,000 contacts is excellent, but the size and the magnitude of your database is just the starting point for creating momentum. To thrive, you must convert such energy into sales and revenue. That's where direct outreach comes in.

Direct outreach means to reach out to your potential clients directly. Just as you would prospect potential customers via phone calls and personalized emails; prospecting and reaching out to new leads is the responsibility of the business development department. To dominate and maximize the conversion ratios, you must perform business development functions to create demand.

Conversion Versus Domination

It is crucial that you change your mindset about conversion. Rather than the typical two-step conversion process of generating leads, and then turning them to customers; your mindset must change! To dominate your marketplace, you must follow a five-step process:

1. Demand Generation
2. Lead Generation
3. Lead Nurturing

4. Lead Conversion
5. Lead Monetization

Amateurs want to start the first step with generating leads. Secondly, and before building rapport or trust, they want to jump right into converting leads to customers. While some agents may get lucky closing a quick sale or two, typically, skipping through multiple phases of the customer's lifecycle results in low conversion ratios, and loss of viable sales. Potential clients will only be ready to commit when you nurture and carefully cultivate them.

While Lead Generation and Lead Nurturing are functions of the Marketing Department, I purposefully did split them into two topics, and therefore two chapters. I want to maintain the prerequisite for nurturing of your database before making your sales calls. Also, I want to set a solid foundation for a successful lead generation strategy using Demand Generation. This way, you leverage your time generating quality leads and contacting only those in the market to buy or sell.

Moreover, while the Sales Department is responsible for Lead Conversion and Lead Monetization, I also divided these two subjects into separate chapters. This way, I can emphasize the importance of qualifying hot opportunities before attempting to turn them to customers. Hence, a real Domination System expands into a five-step process instead of a Conversion System that comprises a two-step process—lead generation and lead conversion.

Implementing the above five steps are highly critical to have a valid business model. They are the foundation for every functional and thriving business—including real estate!

Note

Although Customer Service and Recruitment are outside the scope of this book; you can go beyond sales to leverage customer service where you can influence growth by the quality of services provided. Maximize the growth of your business beyond sales conversion to tap into clients' loyalty and satisfaction. Gain more clients through referrals and partnerships and expand your business through hiring and recruitment.

Take the Next Step

Execute Tactic 8.5: Split Your Domination Business Model into Five Stages

Make the Move

> *"The key to growth is the introduction of higher dimensions of consciousness into our awareness."*
>
> —Lao Tzu

A young man sees a beautiful young woman, walks up to her, introduces himself, and kindly asks for her number so he can ask her out on a date.

A woman walking her dog sees a new family moving into her neighborhood. Respectfully, she will make small talk after she introduces herself. Then, welcomes the neighbors to town and off she goes with her day.

A charming lady makes a phone call to the customer service department to follow up on her online order. She introduces herself, provides the order number to the service representative, and inquires about the delivery date of her item.

Not all young men will end up going on dates. Not all dog walkers will end up being friends with their neighbors, and, not all callers will end up getting the best service they deserve. However, those with interpersonal skills who kindly introduce themselves have a higher chance of making an emotional impact than those who don't.

Top real estate agents have one thing in common! They possess the courage to fire up small talks with strangers, introduce themselves, follow up, and conclude with many listings. Most top agents, their relationships, and colossal commission checks start with the word "Hi."

Veronica Figueroa, Jeff Sibbach, Missy Stagers, Kevin Spina, and Ryan Serhant, among other top leaders in the real estate industry, will advise that success depends on creating and maintaining meaningful relationships.

To build relationships, you must create awareness. To build awareness, you must introduce yourself. Thus, business development begins by presenting yourself and saying "hello" to strangers! Move outside of your comfort zone and make an impact by introducing yourself for higher chances to close more sales.

Introduce Yourself

Since attracting attention is the first step to creating brand awareness, it is necessary to find out where customers are online. It would be almost impossible for Ross to convert people to customers if he doesn't know where they hang out. It would also be challenging to do business with people who do not know him. Hence, there is a need for an excellent introduction to potential customers.

Ross has the following contact information of the people in his database:

- Full Name
- Address
- Mobile Phone Numbers
- Email Addresses
- Social Media Profiles

If Ross wants to find out which channel his potential clients prefer to communicate on, he will need to message them through all possible platforms. The channel with the most responses wins! Each one of Ross's contacts will eventually have a top communication channel. By figuring out how he can communicate more effectively with each member in his database, Ross can ensure his marketing and sales efforts yield the highest impact.

To test the best response rate, you and Ross can use the following communication channels:

1. Offline

    a. Send direct mail to the mailing address on file.
    b. Set up face-to-face appointments, meetings, and events.
    c. Make automated and personalized phone calls.

2. Online

    a. Send text messages (SMS).
    b. Send one-on-one email messages.
    c. Send individual direct messages on social media.

For each contact or cold lead that you haven't reached out to yet, use the Out-Of-Network spreadsheet you created in *Tactic 7.4* to show which channel each contact had responded. It is common for people to respond to several channels. Some may not respond at all. Some may prefer to respond via email, others via text messages or social media. Regardless, the goal is to determine at least one communication channel where you can get a hold of your farm. Whether it is offline or online, your contacts will decide where and how they want you to reach them!

Your First Touch

Your opening message to brand new Out-Of-Network contacts—the 6,000 cold contacts in your database—must be a simple introduction. Create a quick statement to build awareness and inform people in the territory of your presence.

However, to attract the attention of cold prospects, your introduction must entice them. It must be valuable and helpful to your audience. It can't be a pushy sales pitch. Instead of turning people away, you want them to foresee values that separate you from most of the real estate agents within their network or area.

Also, your introduction must be highly targeted and strategically composed for each specific target audience. Based on your farm and the type of people who live in your territory, your email introduction could be as simple as the following example:

"Hi [Sally], my name is [James], I am a real estate advisor in the [Orange County] area. I don't believe we've met, so I thought to make an effort and kindly introduce myself. If you're in the neighborhood, please feel free to stop by our office at [123 Main St. City], or visit my website at [JamesTyler.com]. I look forward to meeting you soon. Thank you!"

I call this a *"Soft Introduction."* Try different variations of your first communication. While the goal is to create awareness, test the response rate for generating interest! Iterate, edit, and optimize your messages until the response rate increases to a minimum of 15% conversion ratio.

It is imperative you reach out to introduce yourself to people that don't know you. Also, to request permission to contact them again before adding them to your Contact Relationship Management (CRM) system, or an email marketing platform.

Before promoting your real estate services to people via CRM, email, or SMS platforms; verify their personal information. Make sure email addresses and phone numbers are valid before automating your introduction messages. If you send communication to invalid data, email and CRM providers, such as Gmail, MailChimp, or Realvolve, may discontinue your services. Avoid bounce backs to stay in good standing with your technology vendors.

You also do not want to face resistance or promote your services to a reluctant audience. Ensure contacts in your database will work with you and require your real estate services within the next 12 months.

After your first touch, you can determine whether you need to remove a contact from your Out-Of-Network spreadsheet, or move them into your CRM for further nurturing and cultivation.

That said, the direct outreach means manually reaching out to your contacts via all available offline and online channels. Direct outreach will create awareness and move contacts from the Out-Of-Network spreadsheet into your CRM.

You may experience some adverse reactions during the direct outreach process and especially the first touch. Some communications, such as email and text messages may irritate people, and a few will bounce. Some individuals will not be interested, and many will not respond! Welcome to the world of business development!

No matter how your communication goes, take nothing personally. Do not argue with anyone! Respond kindly and don't react or get discouraged. You are just introducing yourself! If someone isn't having a great day, they will either not respond or tell you not to contact them. Just keep on trucking! Fortunately, you will come across many pleasant people that will make your day. Also, soon enough, when you figure out what works and start landing new clients, you will forget about all the negatives.

Business development is very similar to sales; it is a numbers game! When someone isn't responsive, move on! Only add people who are interested in your services to your CRM. Once added, tag each segmented group with a proper status. For instance, label them as cold leads, warm leads, or hot leads; also mark those who are past clients, family, friends, or partners. Otherwise, keep on following up and touching your contacts! Remove those who aren't interested or don't respond after ten touches or 60 days, whichever comes first.

> *"Get out of your comfort zone and plant seeds through direct outreach with people you don't know yet. Those are the people you need to serve, and with whom you need to build long-lasting and meaningful relationships."*
> —James Tyler

Although direct outreach may seem like a time-consuming project, it isn't. My Business Development Representative (BDR) reaches out to about one hundred new people each day. You'll see how critical your BD can be once the momentum grows.

While there are multiple options for manually reaching people, I will showcase the ones I've used. It is up to you to choose which method makes the most sense for your business model. I recommend using all the tools I mention in this chapter. Should a tool I have proposed in this book become unavailable, substitute it with a similar technology. Either way, don't stop in reaching out to people and introducing yourself. The power of persistence will do wonders! These individuals will praise you for your efforts, tenacity, and perseverance.

> *"Success in life is a matter, not so much of talent or opportunity as of concentration and perseverance."*
> —Charles William Wendte

Introduce Yourself Via Email

Send free mail-merge emails via Google Sheets and Yet-Another-Mail-Merge (YAMM). It is an add-on to the Chrome browser I highly recommend using to send your initial introductory emails.

Despite that, you have validated and verified the integrity of emails; you will still come across invalid emails once you reach out to people. You want to add only responsive contacts to the CRM to ensure low SPAM and bounce rate, and only when you are 100% confident that their emails are valid. The same goes for mobile phone numbers. Therefore, you can sign up for YAMM and send your emails within the next 30 minutes.

If you don't know how to send emails using Mail-Merge messages with Google Sheets; go to YouTube and search for "Yet-Another-Mail-Merge." There are plenty of videos that will show you how to use this add-on.

Though, not every Gmail account can send to the full official limits, your daily limit is 500 email recipients per day. Also, even though you can send emails to a maximum of 500 recipients per day, Google prefers a maximum of 100 recipients per message using the To, Cc, and Bcc fields of a single email. Thus, do not send all 500 messages at once.

Set up your Gmail account with YAMM to send a hundred introductions every day! Alternatively, you can hire a company like LeadRipple to do this for you. Another option is to use a professional cold email service, such as MailShake, Reply.io, or Prospect.io.

Introduce Yourself Via Text Messages

Send free text messages through the Short Message Service (SMS) using Google Voice. Remember that Google account I asked you to sign up for in chapter 6: SHARPEN YOUR AXE? Great! Log in to Gmail using your personal Gmail account, then visit Google Voice to set up a free account—if you don't have one yet!

Pick a local Google Voice phone number and set it up on your personal or business mobile phone. Also, download Google Voice mobile application from the App Store or Google Play to send text messages using your mobile device.

Use Google Voice to send SMS from online at *https://voice.google.com* or through the mobile app. Send Introduction Messages you compiled in chapter 3: DISCOVER YOUR CORE VALUES.

Moreover, you can also introduce yourself using the Multimedia Message Service (MMS). Here's a quick hack to turn each mobile phone number to an email address (phone-to-email). This way, you can send multimedia messages—with subject, body, and attachments—to prospects' phones from your email address.

This trick is beneficial as it doesn't incur text messaging data and service rates. It is like having a free unlimited texting plan. Also, unlike any other texting application, such as your phone or WhatsApp, by using this method, recipients can't see other parties within the group.

1. Find the carrier for each mobile phone number using FreeCarrierLookUp. (I mentioned many more similar services in chapter 7: BUILD MASSIVE MOMENTUM, under Free Contact Enrichment Resources).

2. Attach the mobile phone number to the carrier's URL. For example, if it is *AT&T* and the phone number is *(123) 456-7890*, then the email would be *1234567890@txt.att.net*.

3. Paste phone-to-email addresses of your prospects in the Bcc field of your Gmail account or any email service provider. Enter a subject line and a short message in the body or your email to send your multimedia introduction messages. Feel free to attach a media file, such as your profile photo or a useful document just like you do with email.

Here are the URLs you'd append to each mobile phone number based on each specific carrier:

- Alltel: *phonenumber@message.alltel.com*
- AT&T: *phonenumber@txt.att.net*
- Nextel: *phonenumber@messaging.nextel.com*
- Sprint: *phonenumber@messaging.sprintpcs.com*
- T-Mobile: *phonenumber@tmomail.net*
- Verizon: *phonenumber@vtext.com*

Do not send MMS messages to large groups. Though some phone service providers allow group multimedia messages up to 10 recipients, I highly recommend against it to avoid having your messages not delivered.

Introduce Yourself Via Social Media

Use social media direct messages such as Facebook Messenger, Instagram Direct Message (DM), and LinkedIn messaging system, or the InMail feature if you are a LinkedIn Premium member to connect and introduce yourself to prospects, and your 6,000 contacts.

Your virtual assistant may reach out to your contacts manually via each social media channel or use automated services, such as FollowLiker or PhantomBuster to speed up the process. I recommend a manual reach out initially until you perfect your introduction message, and then automate once your response ratio surpasses 15%. Remember, the goal is to get a minimum of 15 out of 100 new prospects to engage with you daily regardless of the communication channel.

Introduce Yourself Via Voicemail Drop

Introduce yourself through pre-recorded voicemail messages using companies like SlyBroadcast, DialMyCalls or one of the ringless voice messaging technologies mentioned in chapter 6: SHARPEN YOUR AXE.

You can send thousands of ringless voicemail drop messages directly to your 6,000 contacts in just a few minutes. Voicemail drops can reach cell phones and landlines in both the US and Canada.

Dial-in, record from your phone, or upload an audio message. Upload your contact list to schedule out your voicemails; send them immediately, or in batches. It is that simple!

Introduce Yourself Via Phone Dialer

Upload your list of 6,000 phone numbers to your favorite dialer or one of the dialer services for which you signed up while reading chapter 6: SHARPEN YOUR AXE. Dialers are especially helpful in introducing yourself to contacts with only landline phone numbers.

Introduce Yourself Via US Mail

Send handwritten notes introducing yourself to people with valid mailing addresses verified through CylinderAI, Google Maps, or Google Earth.

Studies show that ink-on-paper notes are more effective than digital messages. Leave a striking impact on your prospects with a hand-stamped envelope in their mailbox. I recommend this outreach method to contacts with valid mailing addresses but no email addresses or mobile phone numbers.

To conclude, here are the top communication channels in the order of their effectiveness at the time of writing this book:

1. Face-to-Face (appointments or speaking events)
2. Text Messaging (SMS or MMS)
3. Direct Social Media Messaging (i.e., Facebook Messenger, Instagram DM, LinkedIn, or Snapchat)
4. Phone Calls (mobile numbers only)
5. Online Engagement (online chat, guest commenting, i.e., social media, YouTube, or blogs)
6. Handwritten Notes (literally, handwritten)
7. Emails

Therefore, you are more likely to get a better response rate by sending handwritten notes than sending emails. Hence, the best two communication channels to focus on are text messaging and social media in reaching out to your contacts.

Take the Next Step

Execute Tactic 8.6: Build Awareness by Introducing Yourself

Execute Tactic 8.7: Create a Top of Mind Content Strategy

CHAPTER NINE

Generate High-Quality Leads

"The vision must be followed by the venture. It is not enough to stare up the steps - we must step up the stairs."
—Vance Havner

In the previous chapter, you learned about the responsibilities of the **Business Development** department. You introduced yourself and created awareness. People in your territory know who you are and what you do. In this chapter, you'll jump into tactics of the **Marketing** department. You want to become omnipresent and be everywhere. Not only promote yourself, but you will also attract attention and raise interest to generate leads. To give you a glimpse into where you're headed, in the next chapter, you will create desire by telling the world the benefits you offer (nurturing leads and relationships). In later chapters, you will use your sales department and sales expertise to create opportunities (lead conversion). Then, you will close sales by converting prospects to deals (lead monetization). To dominate, you will need to excel exceptionally and to be the preeminent leading expert in all these three departments.

When you build awareness and do marketing the right way, people will check you out. If you genuinely understand your audience, give them reliable solutions, you will earn their attention, interest, and trust.

When you sell people on benefits, and they translate your perceived values into a fair price, they will consider keeping in touch and hiring you.

Thus, the marketing department shapes and increases a consumer's perceived value of the real estate services you offer. The sales department breaks down all the benefits to justify your asking price. By marrying the two departments together, and understanding their functions and responsibilities, you are destined to succeed. Your goal in this chapter is to delve into marketing, attract attention, and raise interest so you can generate high-quality leads.

The Blue Car Syndrome Concept

As soon as you buy a car, you see more of your new vehicle; the same make, model, and color everywhere. The number of that kind of vehicle hasn't increased or suddenly changed, but your awareness of it has.

This cognitive bias is a systematic pattern of the frequency and recency illusion. It's nothing more than merely noticing an entity that was always there, but you just had not seen before. That may be because of your little or no experience with that specific car. It's unnoticed in your subconscious mind until you have purchased it and had direct experience with it. Now it has entered your foreground conscious mind, you know of it, and you notice it more frequently.

How do you apply the exact conceptual phenomenon into your real estate business? How do you, as a real estate expert, become the "Blue Car Syndrome" to get people to see you everywhere and all the time? By being omnipresent.

Omnipresence

To be omnipresent—everywhere—you need to target the right audience and deepen the relationships to stay relevant and "top of mind."

The ultimate goal of marketing is to be in the right place at the right time. With omnipresence, that's what you're doing. You are regularly getting yourself in front of your target audience, so you become the only choice for them when it's time to convert.

Omnipresence isn't just about timing, though; it's also about frequency. The more people will see you and your brand, the more likely they'll connect with you. To build rapport and sell more real estate, you don't just appear out of the blue when potential clients are ready to buy or sell. You must show up frequently, consistently offer tremendous value, and, subsequently, build meaningful relationships.

Last, you become omnipresent with the right channel and location. For most intelligent and savvy agents, omnipresence starts with advertising and retargeting on Facebook, but that's just one platform. Instagram, LinkedIn, Google, YouTube, Ad Networks, Email, Text, and push notifications are all channels where you can show up for your audience. Not to discount offline channels—meeting in person, attending network meetings and events, sending direct mail, and placing ads in a local newspaper can also contribute towards your omnipresence.

From a strategic point of view, omnipresence is about concentrating most of your advertising budget on people after they realize you and your brand. This way, you grab more attention, pique more interest, and have a better chance of being relevant to the perfect potential client.

Using the three factors—timing, frequency, and location—of omnipresence in marketing to target your audience creates the "Small Pond, Big Fish" effect. Essentially, you can quickly become the expert inside your niche—territory or your farm area—in what seems like overnight. The premise of most things online is always to have "more, more, more," but omnipresence is about depth.

Consistently be everywhere and become top of mind by being in front of your audience at the right time, with the right frequency, and on the right platforms. It would almost be impossible for your potential clients not to see you everywhere, every single day. Omnipresence is a critical element for attracting attention and raising interest. Hence, the secret to consistent lead generation!

> *"We must strengthen the bridges that connect us with others; the shorter the bridge, the closer they are, and the stronger the relationship."*
> —James Tyler

Lead Generation

Lead generation isn't about collecting people's emails and phone numbers. It is about getting people interested in what you do—real estate. Good leads are people buying or selling in the next 12 to 24 months. Otherwise, they are just glorified contacts in your network and database.

In chapter 8: PLANT MORE SEEDS, you learned about demand generation. You have reached out and created awareness with 6,000 contacts. Now, it's time to sift through them and filter out potential leads that might turn into delightful clients.

People consist of different demographics. Most will hang out often online, and others might only be available through offline channels. You must be omnipresent to determine their intention and find those who are interested in buying or selling. Therefore, you can generate leads from two primary sources: Offline and online channels.

Offline Lead Generation

The offline lead generation channel is dying out as quickly as generations evolve. I would guess we still have another decade before offline lead generation tactics become obsolete. If you live in high-tech areas where innovators, early adopters, millennials and Generation Z live, then you are better off generating leads from the online channels only. However, if your territory is predominantly baby boomers[i]—or Generation X[ii]—the offline lead generation channel might be more effective than the online channel. As a good rule of thumb, make small investments in both channels to find out what works best in your specific area and for your target demographic.

Generate Leads through Local Branding

Your goal is to get involved with your local community and become a celebrity agent, a local authority, and a local influencer. Generate personal brand awareness by increasing people's familiarity with you and your real estate brand. Provide convenient methods of access to you as a person. Be available online, via phone, email, and social media. Also, stay active within your community, network with people at parks, local stores, and local events. Face-to-face conversations are far more powerful than online channels.

Generate leads by building rapport and a well-known name not only for yourself but also for your real estate brand. Introduce yourself, who you are, and what you do for your local community. Be transparent with people and local brands, so they become familiar with your name, your face, your logo, and your contact information.

You will encounter less resistance and more likely to be the agent of choice when you get involved with residents and businesses consistently. You want to be top of mind when the time is right. Don't limit your promotions to homeowners only. Business owners are potential home buyers and sellers too. The more frequently they see your brand, the more likely they will remember you. Be in their face online, in person,

i   Birth years ranging from the early to mid-1940s and ending birth years ranging from 1960 to 1964.
ii   Birth years from the early-to-mid 1960s to the early 1980s.

in their mailbox, and at local events. Earn more trust and quality leads by showing your face to local companies regularly.

You don't just sell houses; you sell towns! Become familiar with every street and corner in your area. Make sure every person at every community, attraction, company, store, and ice cream stand knows who you are. Lead generation outcomes from local branding will vary based on your specific neighborhood. Test different strategies in your territory and determine which methods work best in generating the best leads.

For example, if you live in the city, you can organize regular walking tours of historic neighborhoods for your sphere. Hire a professional guide for the events so you can bring more people to attend each trip. Talk about the history and the benefits of living in particular communities or neighborhoods. If you live in the suburb, host an event at a local restaurant that shows off your love and expert knowledge about your farm area. You can showcase the best characteristics that your city offers from landmarks to farmlands, businesses to schools, and similar sites.

Generate leads from local branding by attending City Council meetings, HOA meetings, and local events or by interviewing owners and managers at local brands. For example, visit food chains such as restaurants, coffee shops, donut shops, grocery stores, and juicers. Other options are financial institutions, such as banks and credit unions. Also, government officials such as the mayor or county executives, and medical centers, such as hospitals and emergency clinics. You can also generate quality leads by partnering up with local stores, companies, and service providers—for example, gyms, nutrition shops, flower shops, print companies, alarm system companies, attorneys, carpet cleaning companies, contractors, home inspectors, furniture stores, home remodeling companies, insurance agents, interior designers, lawn maintenance and landscaping, mortgage lenders, home stagers, and more. Leverage partnerships with local brands to become the go-to real estate advisor. Your options are vast!

> *Take the Next Step*
> *Execute Tactic 9.0: Generate Leads through Local Branding*

Generate Leads through EDDM Farming

USPS® Every Door Direct Mail® (EDDM®) is a common offline lead generation technique used by most real estate agents. It is an affordable targeted advertising technique that lets you map your marketing mail audience by age, income, or household size. You can use the EDDM mapping tool to choose a zip code and a carrier route that will target your best customers. For example, you can target residential areas with people between 35-54 years of age and an annual income that is higher than $100,000.

With an average of $0.187 per piece, you can deliver individual direct mail campaigns up to 5,000 addresses. Each mailpiece size can be 10.5" by 6.125" or higher. That is plenty of room on one marketing piece to promote your brand.

Use one of the printing partners you chose in the "Local Branding" strategy. Alternatively, "Find a Printer" in your local area right on USPS's website to handle the design and printing of your EDDM flyers. However, instead of paying full price for your EDDM farming campaigns, you can use sponsorship through local partners. This way, you and your partners will reach more audience at a lesser shared investment.

Attract Home Sellers Using EDDM

In chapter 3: DISCOVER YOUR CORE VALUES, tactic 3.9, I shared with you a unique value called "Certified Pre-Owned Home (CPOH) Program," which comprises a four-step process:

1. Home Warranty
2. Home Inspection
3. Pest Inspection
4. Appraisal

Suppose the mailer of an EDDM campaign to 5,000 residents will cost you $1,000. Instead of paying the full amount of $1,000 for the marketing campaign, you can invite those four service providers mentioned in the CPOH program to take part in co-branding with you on the mailer. One full page of your promotional material will display your real estate services. The other page will advertise the services of your four partners—home warranty, home inspection, pest control, and an appraiser.

Split one side of your marketing piece into four quadrants. Each partner contributes 25% of the overall cost of the mailer—in our case, that is $250. Each partner will get an ad on the back of your postcard. You can offer to pay for design and print facilitation costs of the mailer—which shouldn't cost more than $250. Get four partners approvals for the design and send a direct mail campaign.

By following this strategy, you would save $1,000. Instead of paying

$1,250 to reach 5,000 potential sellers, you would only pay $250. Clients will also view you as a credible real estate professional since you have highlighted four trustworthy partners on the same marketing piece. Your partners send the advertisement to all 5,000 residents at nearly 25% off the overall cost. All five companies can reach 5x more homes and therefore generate 5x more potential clients. Instead of only reaching 5,000 homes at $1,250, you can now contact about 50,000 homes at $1,250.

Attract Buyers Using EDDM

You guessed it. Repeat the same process above using EDDM to gain potential home buyer leads. You can partner with several companies that buyers can hire when representing them. Connect with a preferred lender, a mortgage broker, an appraiser, a law office, escrow, a title company, a moving service, or a home security system provider.

Top real estate professionals do not knock doors neither do they host open houses. However, most of them still farm using offline marketing tactics to support their online lead generation efforts, such as co-branding and EDDM. By leveraging the shared-cost marketing strategy, you can do what the big guys are doing, but five times the scale, and at a much cheaper cost.

> *Take the Next Step*
>
> *Execute Tactic 9.1: Generate Leads through Every Door Direct Mail (EDDM)*

Generate Leads through Publicity

Publicity, or Public Relations (PR), could be a game-changing lead generation channel. Getting press coverage remains the most effective channel in building brand awareness and generating an immense amount of leads. Publicity heightens credibility and trust factors, hence leading to quicker sales and revenue.

Quickly uncover lead generation opportunities by differentiating yourself from the competition. Take advantage of this strategy, as most real estate agents tend to pursue other marketing tactics to generate leads. PR is underutilized as a tool for customer acquisition. It could be because of the several moving parts of a publicity strategy, the high costs of PR agencies, and the difficulties of getting into the press.

You need to be the PR Firm and the celebrity that will thrive and dominate your real estate market. You don't need a PR firm and tens of thousands of dollars to get publicity coverage. All you need is a story, a press release, and a Rolodex of media journalists.

Whether you want to be featured on the cover of your local magazine, on a TV program, or in a newspaper; the process is the same. You must write a compelling story, or a press release, and pitch it to influencers.

While a press release aims to get the attention of reporters, the ultimate goal is to build awareness and generate quality leads. Before having influencers get your message to your target audience at a scale, be equipped to move your audience to a landing page where you capture new subscribers.

The main recipe of a successful PR campaign is a compelling story that is remarkable and newsworthy! What story can you write that is worthy of getting published in the mainstream media? While it can, the story doesn't have to be about you and your business. Journalists ignore self-promoting pitches. To get invited to a radio or a TV program, or published in high-tier media outlets like Forbes and Inman, you need to write stories that will benefit your audience and the media. Your business will eventually reap the rewards, and people will reciprocate once they recognize your values.

Write a story about trending news in your local area and how that can affect the local real estate market. Write about a new development in your community and how it can help solve an inventory issue that home buyers have been facing. Write about the recent real estate market changes in your county or state. Always support your story with viable data, charts, and infographics. Write about topics you are most competent with to increase your chances of getting published. Thus, increasing your credibility and build enduring authority when media outlets reference your name and link back to your website!

Generate media hits with the outlets that matter most for your brand and your story. Build targeted media outlet lists of top tier publications such as Inman, New York Times, Wall Street Journal, and CNN. Also, create a list of mid-tier blogs and smaller local newspapers. It is imperative to pick media outlets that would publish your story. For example, The Los Angeles Times is more suited to post about a local storyline in Los Angeles city than The New York Times.

Once you've chosen the proper outlet for your story, you can then research the journalists who write about real estate to cover your article. You can use the outlet's website or Twitter Advanced Search to find the contact information of specific real estate journalists and reporters.

Build a comprehensive social influencer database and media contacts. Create lists based on specific topics, coverage, market, industry, and media types. Find contacts at traditional print publications, like newspapers, magazines, and journals; also, television, radio, newswires, digital or online outlets, and blogs. Ensure sorting and categorizing of contacts so you can narrow your segments down to journalists, reporters, editors, publishers, contributors, columnists, and freelancers you need to reach.

When you write articles that provide value to your local community, journalists may wield some of its information and reference data you've used to link back to your biography or website. In other cases, they may feature your entire content. Find out what journalists are interested in writing about before you pitch to them. Find those who write articles similar to your story. Then, provide value by emailing them a few useful columns with relevant subjects. Once you build a good rapport with the journalist(s), then you can pitch your story. This way, you will have a much better chance fully featuring your paper.

When approaching media outlets about your story, journalists will most likely read your personalized one-on-one email. Do not blast an email to hundreds of influencers—commonly known as the shotgun pitch or focused pitch. Send authentic and customized emails that are short and simple. Get to the point and do not sound desperate.

Most give up too early in the publicity strategy. If you don't hear from your journalists, follow up within 5-7 days after your first email. Then, reach out by phone to find out what would motivate them. Often it is more about timing than your actual story. Be resilient, consistent, and persistent. The process gets easier once you get your first release published. The first one is the hardest. Don't give up!

Use one of these management platforms to power your PR communication strategy in a unified way across all channels: Buzzstream, Cision, Newswire, Meltwater, Prowly, Prezly, pr.co, and Muck Rack.

Take the Next Step
Execute Tactic 9.2: Generate Leads through Publicity

Generate Leads through Phone Prospecting

You have 6,000 contacts in your Out-of-Network database you gathered in chapter 7: BUILD MASSIVE MOMENTUM. In chapter 6: SHARPEN YOUR AXE, I shared with you information about lead sources such as FSBOs and Expired listings. I also shared with you a few resources about dialer systems. With the 6,000 contacts in your database—with valid and verified mobile and landline phone numbers—you can put your dialer into action, and get to speak with thousands of people.

Use the Voicemail Drop tactic to drop pre-recorded voicemails to every contact of those 6,000 people directing them to a pre-recorded voicemail. Your pre-recorded voicemail needs to forward callbacks to a pre-recorded message and a tracked phone number to measure conversions. You can call back people who show an interest in learning more about your services. This way, you would only speak to those who are in the market of buying or selling in the next 12 months.

If you want to attract sellers, you can leave a voicemail about a few qualified buyers who are interested in buying. Also, you can run another campaign asking your target audience if they know anyone in the market to sell. On the contrary, you can leave voicemails about new listings in the area and if your target market knows anyone interested in buying. Test different messages, measure the response rate, optimize, and re-run your marketing campaigns.

There are other offline lead generation tactics you can consider. These include but are not limited to:

- Published Print Ads
- Paid Radio and TV Ads
- Seminars and Corporate Events
- Sponsoring Events
- Hosting Events
- Outbound Sales Reach Outs
- Promotional Branded Goods (car wraps, physical flyers, brochures, and handouts)
- Networking with People at Events
- Travel to Conferences & Trade shows

No matter what strategy you choose for your lead generation, make sure you track key indicators to measure the performance of each campaign. Test with a small group of people, measure, optimize, and resend your campaign to bigger groups. Iterate the process until you have a perfect offline lead generation system.

Take the Next Step

Execute Tactic 9.3: Generate Leads through Automated Phone Prospecting

Online Lead Generation

With the proliferation of various technologies, such as the Internet of Things (IoT), and mobile applications, it has become clear that online lead generation is a must-have for every business. There are plenty of marketing tactics that you can use to broaden your reach besides local branding, direct mail, local newspaper, and prospecting.

Nowadays, the most common online lead generation method is social media advertising. However, before we explore the different options of online lead generation, let us ensure that you have what you need to gain internet leads. Check to see if you have met these seven steps that will prepare you to generate high-quality leads online:

1. Did you create your branded content mentioned in Exercises one and two from chapter 5: SPEAK YOUR MESSAGE WITH INFLUENTIAL CONTENT?

2. Did you set up your blog, your business website, an IDX solution, and the lead capture landing pages mentioned in chapter 6: SHARPEN YOUR AXE?

3. Does your blog or business website show your real estate services, guarantees, testimonials, reviews, and all values described in chapter 3: DISCOVER YOUR CORE VALUES?

4. Did you add your UVP, biography, executive summary, and your story to your websites as mentioned in chapter 4: ELEVATE THE POWER OF YOU?

5. Have you collected and organized all data for the 6,000 contacts of Sellers, Buyers, Foreclosures, FSBOs, and Expired Listings mentioned in chapter 7: BUILD MASSIVE MOMENTUM?

6. Did you plant plenty of seeds by introducing yourself to those 6,000 contacts, as mentioned in chapter 8: PLANT YOUR SEEDS?

7. Does this strategy seem to align with the short-term goal you set up in chapter 2: ARTICULATE DESIRES INTO TANGIBLE GOALS?

If you are a legacy leader with a growth mindset, you would have all seven items accomplished already. Thanks for trusting in me and for following through this far.

For those who have a fixed mindset—yes, those who haven't finished all seven items that I just mentioned above—stop reading or listening! Consuming tons of information without taking action will not benefit you in generating online leads. You can either go back to buying leads from Zillow or finish all seven items before proceeding. Your choice determines your online lead generation faith!

Take the Next Step

Execute Tactic 9.4: Prepare for Online Lead Generation

Generate Leads through Content Marketing and Blogging

Content marketing is the foundation of growth. If you get content marketing working, then you can re-target people. You can create lookalike audiences on different channels. If you're creating great content, it builds links, which brings your domain authority up, which generates more traffic, more leads, and hence, more sales. You can write more content, then collect more emails, and then optimize your conversion rate from there, but everything starts with content first.

Your websites are the backbone of your business. If done well, they can impress your visitors and generate quality online leads. Your blog is an excellent resource for driving traffic. Blogs are generally rich in content, so they are invaluable for building online brands. Your business website builds trust and credibility. IDX integration serves your potential clients' buying and selling needs. While you can host these sites on one domain or website, keep in mind, they serve different purposes. All of them together are the recipe for generating great leads.

Remember, people browsing your blog, or searching your IDX website may not be ready to buy or sell immediately. They may just be researching. Often, they don't even know what they want. Aim to educate your audience with valuable content. Offer less ambitious calls-to-action (CTAs) and be mindful of pushy and "salesy" copy. Ask your readers to subscribe to your newsletter, or to follow you on social channels.

It is okay if you don't get people's contact information within their first visits to your sites. Just them being on your website, you are now tracking them with Google, Facebook, and LinkedIn pixels. You will later generate leads by retargeting them through paid ads. Again, if you haven't installed the remarketing pixels, please follow the instructions given in chapter 6: SHARPEN YOUR AXE under the section titled "Sharpen Online Tracking." Using content and landing pages is by far, the best strategy in generating the highest quality of leads online.

> *Take the Next Step*
>
> *Execute Tactic 9.5: Generate Leads through Content Marketing and Blogging*

Following the steps provided in *Tactic 9.5* will create the *"Blue Car Syndrome"* effect. This content strategy will drive thousands of people to your websites and hence your landing pages. You will capture hundreds of leads. Your CRM will pull their contact information from the landing page automatically. Also, seamlessly sync the data with ContactPlus and your mobile phone. (Visit chapter 8: PLANT YOUR SEEDS if this process doesn't ring true to you). Otherwise, ensure that you have a CRM that sends you mobile notifications as soon as you receive an online lead.

Now, all leads are at the convenience and the palm of your hand. They are just waiting for you to follow up with them and turn them into loyal customers.

While lead generation tactics may differ, the concept is usually the same; get them from everywhere to your world—your websites. Then provide them with valuable information—content—in exchange for their contact information so you can follow up, build relationships, and turn them into advocates.

With this content marketing strategy, you would typically drive traffic to a landing page where readers would download useful documents in exchange for their email address. Offer valuable content such as eBooks, white papers, and real estate guides. Later, you will use each subscriber's email address to nurture via email marketing, turning them to loyal clients.

Generate Leads through Push Notifications

A recent survey from comScore states that over half of 91.6 million unique visitors to all the real estate sites were from mobile devices. Another REALTOR® survey in 2018 says that 85% of all buyers used a mobile device during their home buying process.

You can generate leads using email messages, text messages, and social media messages. You can also do so by sending push campaigns. Only a few agents take advantage of push alerts. Though, there are only two types of push notifications that you can leverage: web and mobile.

Web Push Notifications

Web messages—also known as web push notifications—are push

notifications sent to a user via desktop or mobile web browsers. You can activate web notifications by installing a small script on your websites. Users subscribe once they "Allow" notifications from their web browser upon visiting your site. Web messages appear on the screen even when a user is not active on the website, offering increased brand visibility and deeper relationships with web visitors.

Web notifications allow you to generate traffic and leads. Also, engage with web visitors without having to ask for an email or a phone number. One can use them for promotional, educational, and transactional purposes, not just lead generation. However, to build trust and provide utility, it would be best to start with notifications that are transactional and educational. Establish better relationships by showing value and allowing a safe transition to promotional content.

Web notifications are a great way to share educational information such as real-time market updates, available inventory, local events, or trending news. You can mix it up by sending promotional notifications of a listing you just picked up, an upcoming open house or a home price reduction.

Learn about your web visitors and what they look for to provide a more personalized experience with your real estate brand. Tag their preferences and deliver web notifications that resonate with them. For example, send alerts about a new listing in their favorite location or community. Dive even deeper to learn about their home preferences. How many bedrooms and bathrooms are they seeking? Would they prefer a house with a pool? How about the number of stories, the size of the garage, or backyard? When relevant opportunities arise and a new "3-bedroom house in Greenland Community Just Listed," you can use such information to automate the sending of web notifications to interested buyers. Provide value and increase utility to visitors by being relevant. Allow them to build brand affinity and take favorable actions by being served right there and then.

Web notification platforms provide you with subscribers' geographical information such as time zone, the local language, country, IP address, browser name, device version, and type (i.e., desktop versus mobile). Use such information to discover subscribers' areas of interest. The location is a good sign that is the area where they are interested in either buying or selling.

Take the Next Step
Execute Tactic 9.6: Generate Leads through Web Push Notifications

Mobile Push Notifications

"Website audiences are almost 3x the size—and growing twice as fast as app audiences, but people spend one-twentieth the time they do on apps."
—comScore's 2016 U.S. Mobile App Report

Since people spend more time on mobile than the web, it would only make sense to focus most of our online lead generation efforts on mobile—and the internet. A mobile push notification is an alert message that pops up on a mobile device. Push notifications are just like SMS text messages and mobile alerts, but they only reach users who have installed your mobile app. Each mobile platform has support for push notifications—iOS, Android, Fire OS, Windows, and BlackBerry. They all have their own Operating System Push Notification Services (OSPNS). Thus, it is imperative to have a mobile app that works across all mobile operating systems.

Generate Leads Using Mobile Apps

Mobile app publishers can send push notifications freely and whenever. Users don't have to log into the mobile app or use their devices to receive them. Just like web notifications, publishers use them to send transactional and promotional messages. The goal is to get users to take immediate action on each announcement.

Though you can use push notifications to improve engagement and customer experience, in this section, the focus would be on promoting your real estate services to increase sales. Convert unknown app users to known customers by driving mobile subscribers to landing pages such as downloading a guide, or a market report, and other marketing channels, such as social networks.

Push notifications can also become activate when users visit an app store, download, and then install mobile apps. Users enable them once they enable notifications on their mobile devices. Use RPR, SmartAgent, or HomeStack if you don't have an app that captures subscribers. For more details, see "Mobile Apps" in chapter 6: SHARPEN YOUR AXE.

Push notifications are a direct path of communication with users. App publishers should treat the ability to communicate with users via push notifications as a privilege, not a right. App publishers must provide value; if they don't, users will ignore push notifications or turn them off. Some people will uninstall the mobile app altogether.

Generate Leads Using Beacons

Using a beacon is another way to gather mobile users and send push notifications to turn subscribers into customers. It is a hardware device that can emit and receive Bluetooth Low Energy (BLE) signals. Beacons are low-power devices that communicate using the new Bluetooth Core Specification Version 4.0.

Opting in through beacons on iOS apps requires a user to grant permission for an app to send them push notifications, while Android and Fire OS do not. Convincing users to opt-in is vital for the success of apps on iOS, especially since there are 63% iOS vs. 37% Android users in the US, according to DeviceAtlas 2018 Stats.

Leverage beacon technology in attracting leads from local areas such as your office, open house events, movies, malls, and coffee shops. Agents can travel with them, lend them to sellers, or put them on their listings. They can transmit virtual tours to people who drive by, or send interested buyers customized short 40-character messages with a single property website link. Users will then receive alerts on their smartphones instructing them to swipe the alert to see the listing. Users can then open up the agent's app if they have already downloaded it, or download it from the app store to see the in-app notification message.

Personalize push notifications to a group of users or target a specific segment of your user base. Doing so gives you a significant advantage when compared to personalized email and SMS text messages. However, similar to email and text, they also require the management of user identification data such as names and mobile phone numbers. Also, they need an interface for writing messages and sending them. Fortunately, you have complete control of the content displayed through them and the app through a back-end system. Use one of the push notification platforms mentioned in chapter 6: SHARPEN YOUR AXE to capture mobile-based clients.

Beacons have a range of up to 150 feet, or around a 100-yard radius. Some manufacturers allow users to adjust broadcasting power, which directly affects the expanse of the beacon. They come in different shapes, sizes, and costs. The price range is from $5 to $30 per item. Here are a few services where you can purchase your beacons and capture leads using push notifications: BeaconStac by MobStac, SimpleBeacon, BeaconInside, BeaconSage, Estimote, Accent-Systems, and BlueCats.

Apple has divided the range of beacons into four proximity states:

1. **Immediate**: The device is physically close to the beacon and is probably being held directly next to it (within a few inches).

2. **Near**: With a clear line of sight from the device to the beacon. It would show the proximity of approximately 10 feet.

3. **Far**: This state shows that the device detected the beacon, but the accuracy of the proximity is too low to determine if it is Near or Immediate (over 30 feet away).

4. **Unknown**: This state shows that the device cannot determine the proximity of the beacon. Unable to detect it because the ranging has just begun, or that there are insufficient measurements to determine the state (out of range).

You need to measure and test web and mobile push notification strategies. Tactics such as maximizing opt-in rates, generating leads, ensuring new users are properly on-boarded, and reducing user churn rates are all critical to an app's success.

With these web and mobile push notification strategies, you would drive traffic to landing pages where readers would subscribe to get notification alerts. In exchange, they will provide their mobile phone numbers. Later, you can use each subscriber's phone number to nurture via SMS text messaging marketing campaigns and use your favorite sales dialer system to turn them into clients.

> *Take the Next Step*
>
> *Execute Tactic 9.7: Generate Leads through Mobile Push Notifications*

Generate Leads through Online Chat Applications

When you understand how people like to engage, you will find out that most would prefer the chat over email and SMS text messaging. People walk into different stores at the mall shopping around for specific items with ease. That is because the sale representative isn't waiting at the door asking every shopper coming into their store for their email address or phone number. Otherwise, you'll see fewer bodies going into those stores. If you haven't built rapport, people are less likely to provide their personal information.

Shopping online is no different. People leave websites they browse when windows pop up everywhere asking them for their email address and their phone number. Capture visitors' attention and ease prospects in by using online chat widgets. They have become one of the leading website lead generation tools—especially when the chat widget doesn't ask for an email address right away.

Customers want to see the value first before providing their contact information. The purpose of the chat widget is to ease customers in and take micro-commitments. First, get them to chat with you. Then ask questions to let them learn more about your services. Answer their needs and service them well. Once you've get their attention and build rapport, then ask for their email address to send them a valuable document. Once you show value, they will provide you with their phone number to continue the conversation. Otherwise, ask for their digits after you exchange a few emails. This way, you come off less of a salesperson and more of an advisor—and someone that cares by having a meaningful conversation, and personalized one-on-one customer service.

Chat is a powerful tactic in generating quality leads. If used properly, it can be the only online intake form that your business needs to generate internet leads. Via chat, you can target online visitors to find your sales-ready leads instantly. With the chat widget and chatbots, you can automate lead qualification, arrange next steps, schedule meetings, and sync customer data to your CRM.

Install the live chat widget on every website; your business website, blog, landing pages, and single property websites. Sign up for one of the chat widgets mentioned in the "Chat & Chatbots" section in chapter 6: SHARPEN YOUR AXE. Take advantage of conversational marketing for generating leads, whether you're at the office or on your mobile device. Chat and chatbots will allow you to reach online visitors anytime and anywhere.

> *Take the Next Step*
> *Execute Tactic 9.8: Generate Leads through Online Chat Applications*

Generate Leads through Social Media

Using social media platforms, you can generate leads in four ways:

1. **Organically**: through consistently and regularly posting valuable content to all social media platforms such as YouTube, LinkedIn Pulse, Medium by Twitter, Instagram Stories, and IGTV
2. **Paid Ads:** through paid ads by interrupting your audience and buying your way to promote your brand through all channels such as Feeds and Stories
3. **Social Communities:** Facebook and LinkedIn groups.
4. **Private Messaging:** Facebook Messenger and mobile apps like WhatsApp

Generate Leads through Organic Social Media Content

When sharing content online, you want to provide valuable information your target audience would want to consume. When writing a blog article, recording a video, or creating a social media post, think about how the information you are sharing would benefit your customers.

For example, potential sellers would want to see real estate agents share information about market trends and home prices. They're thinking of selling, and they are waiting for the right time to sell. If your social media post reveals a market report showing "now" could be the best time to list, there is an excellent chance they will hire you. The purpose of your online content is to educate and allow your audience the opportunity to contact you, thus generating quality leads.

If you are a selling agent, then it would behoove you to share market data about local inventories, school ratings, and the benefits of living in particular communities. Such information will not only attract potential buyers but will also increase your credibility as a trusted local real estate advisor tremendously. Hence, generating quality buyers leads.

It is essential that you create valuable and compelling content to which your audience will gravitate. Aligning your core values and the mission of your business, combined with your ability to write enticing content is the formula for a successful social media lead generation strategy.

The most cost-effective tactic in generating leads without having to pay for ads is by combining the values and content your audience seeks. Distributing such crucial information to potential buyers and sellers through social media channels is easy. Nevertheless, it is sad to say, but most agents do not use social media the proper way.

Instead of using social media channels to carry meaningful conversations and build relationships, most use them to self-promote their business. Moving forward, whenever you share content to social media, ask yourself: "Is what I am sharing worth talking about?" If the content doesn't warrant a conversation with family, friends, and colleagues, it is most likely noise, and not worth your time neither your audience.

Stand out from the crowd and share information that isn't publicly or readily accessible. For example, inventory information from MLS, charts, and stats from professional services such as Altos Research. Curate real estate news that matter to buyers and sellers. Publish real estate buying and selling guides. Share your opinion about where the real estate market is headed. Educate your market with downloadable eBooks and white papers. Write blog articles about communities, neighborhoods, schools, and local businesses in your territory. Saturating your brand with such valuable content will rank you among top real estate advisors in the nation. Then, when you share your authentic knowledge on social media platforms, your business will thrive.

That said, social media platforms, feeds, stories, and live videos are not only meant to self-promote. You generate leads when you provide value, not when you brag about your real estate services. Good practice or a general rule of thumb is to share seven valuable posts to a single promotional content.

> ### Take the Next Step
>
> *Execute Tactic 9.9: Generate Leads through Organic Social Media Channels*

Notice, in adhering to the steps provided in *Tactic 9.9*, you don't directly share your blog article on any of the social media feeds. This method is quite the opposite of what most people do on social media. Instead, you will only use the social media feed to notify your network about the valuable information you wrote or recorded. Shifting your audience's perception of value from making your brand message about you, to more about your customers, is what makes and breaks lead generation strategies.

By sharing worthy content on LinkedIn Pulse, Medium, YouTube, Instagram Story, Facebook Story, and Facebook Live, you are creating a vigorous momentum. It will continuously drive traffic from the top social media platforms to your blog or landing pages. Also, combined with the social media traffic, repurpose your content to YouTube videos, and watch how Google will reward you by sending you even more traffic from organic search.

Instead of boasting on social media feeds, you are one of the few who understand lead generation. Adding value through organic social media sharing is powerful. Best of all, it is free.

Generate Leads through Paid Social Media Ads

To create high-impact social media advertisements and run successful real estate marketing campaigns, one must create ads that grab people's attention. Written or said advertisement's copy must pique interest, promote responses, entice users into taking action, and move prospects closer to the buying decision. Finally, placing your ads on the right social media channel to reach the right audience is fundamental.

Cookies

When I said cookies, I didn't mean the type of cookies we eat. I meant pixels. The kind Google and Facebook bake daily. People are usually on Google, YouTube, Facebook, Instagram, and LinkedIn. Since you can't physically reach everyone daily, you will leverage technology to do so digitally online. You will use pixels for retargeting your audience and reminding them of you and your real estate services. If you haven't installed Google and Facebook pixels on your websites, please do so now. For more details, see the "Web and Social" section under "Sharpen Online Tracking" in chapter 6: SHARPEN YOUR AXE.

That said, make sure you have Google, YouTube, Facebook (including Instagram), and LinkedIn pixels installed on all your websites. These social media channels will follow your web visitors around on the web, showing them your promotions and advertisements to bring them back to your landing pages. Then, you can capture their personal information and turn them into clients.

Headlines

Just like I iterated the importance of great headlines in chapter 5: SPEAK YOUR MESSAGE WITH INFLUENTIAL CONTENT, real estate advertisements must have captivating titles. Research shows that over 80% of people scan through captions. Only the most compelling headings tempt online users to read further down the ad copy. You can write amazing ad copy, but if your headline isn't catchy, 80% of the people will pass right over your ad.

Ad Copy

Contrary to headlines, writing persuasive ad copy in the body of your advertisements is essential. Let readers imagine what it would be like to live in their dream home, emphasizing on benefits instead of only features. For example, instead of stating that a home has a large backyard, inspire readers with statements that trigger an emotional connection. *"Enjoy an afternoon barbecue in the spacious patio while watching your kids play in the enormous backyard."*

Besides stirring emotions, evoke interest by turning great features into descriptive owner experiences. For instance, instead of conveying a house with an air conditioner, tap into the benefits of coolness and comfort! *"Enjoy the tranquil feeling of a peaceful good night's sleep in an air-conditioned home!"*

Run and test various lead generation ads on Facebook Audience Network, Facebook Feed, Facebook Stories, Facebook Groups, Facebook Messenger, Instagram Feed, Instagram Stories, LinkedIn Feed, LinkedIn Groups, LinkedIn InMail, YouTube, Google Search Ads, and Google Ad Networks.

Create and test various ad types. Not every ad has to be "download an eBook" or "get your home value." Use advertising strategies to increase your awareness and credibility, such as once-a-month social proof advertisements or customers' success-story video ads. They also contribute to your lead generation strategy. Be creative and experience various types of ads. For example, attract sellers with targeted campaigns such as once a week expired listing ads.

Call to Action

What distinguishes great ads from good ads is the ability to build a sense of urgency and to prompt people to take immediate action. Every time you create an ad, give readers a reason to act and concise instructions on how to take the next step.

Persuade your audience by crafting clear calls to action that convey benefits, scarcity, and exclusivity. Use low inventory levels or rising interest rates to urge quick phone calls. Spur people into taking action sooner than later. For instance, to induce excitement, and drive people to a home showing appointment, describe the exclusive nature of your listing and convey that this one-of-a-kind home comes onto the market rarely, and moves quickly.

You must measure advertisement cost and performance by *Click through Ratio* (CTR). The more people click, the better your ads will perform. Social media platforms such as YouTube, Facebook, Instagram, and LinkedIn will reward you if your ads have high CTR. They will show your ad to far more people and far more often than competing real estate agents. The more clicks your ads get, the more money advertisers make, and the more leads you gain. To keep you hooked, they will also reduce ad costs over time so you can keep your ads active. It is a win-win situation.

That said, to increase CTR, measure your ads by clicks first, then won leads. To get your audience to click, I want you to remember this while crafting your call to action: Instead of readers asking themselves, *"Should I click?"* You need to get them to ask themselves, *"Am I A or B?"*

Alternatively, *"Which option applies to me?"*

For example, creating call-to-action buttons such as *"Click Here,"* *"Learn More,"* *"Apply Now,"* and *"Get Your Home Value"* put the reader in the frame of mind of *"Should I click?"* Hence, most choose not to click. Instead, calls to action such as *"Are you a buyer?"* and *"Are you a seller?"* or *"Looking to buy?"* and *"Looking to sell?"* put the reader in the frame of mind of *"Which situation applies to me?"* Thus, most click on whatever applies to them.

Just making that little shift of evoking interest instead of telling through calls to action, helps create significant improvements in Click-Through Ratio. A higher CTR means better performance in the paid ad results and more leads.

Choose the Right Social Media Channel

To create an ad copy that generates quality leads, you must reach the right audience on the right social media channel. Otherwise, you will waste your real estate marketing efforts on people who aren't interested, neither capable of buying your offering.

For example, suppose that you have a listing for a home near a golf course. If you place your ad on Snapchat that predominantly reaches young generation, you can bank on a little to no response to your efforts. However, if you test the placement on LinkedIn targeting people who are interested in anything related to golf, your marketing message will immediately reach potential buyers.

Don't advertise blindly on a particular social media channel because everyone else is doing it. Just because others are promoting there, doesn't mean it is the right move for you or your specific listing. In the example above, you might be tempted to advertise on Facebook and Instagram when, in fact, LinkedIn is a better choice. While potential buyers may be everywhere online, golfers are more likely to hang out on LinkedIn rather than Facebook and Instagram. Even if they aren't interested in your listing, someone in their network might be.

Test Click-Through Ratios and leads received from each ad set, ad campaign, and each channel. Test ads on the channel level—for example, YouTube, Facebook, Instagram, LinkedIn, and even Pinterest. Also, test ads on placement levels—for example, the Feed, Messenger, and Stories. Then, optimize your ads for better results and more leads.

Measure the performance of your ad campaigns daily and maximum weekly. Place the best performing ads on the best performing social media channels so you can get the best returns on your social media marketing investments. Test, measure, and optimize. Rinse and repeat.

Choosing the Right Audience

The key to striking ad placements is to know your target audience. It is imperative to segment your target audience based on your goal requirements aligned with your prospect profiles. Choose the biography of the target prospect you're trying to reach, including geographic location. Demographic facts such as age, gender, ethnicity, income level, education level, marital status, household size, and other data. Also, lifestyle characteristics, including personal interests, activities, and behavioral patterns.

Consider advertising to smaller targeted groups than an extensive network. Dig deep rather than going wide. For example, instead of marketing to a blanket of an audience in a specific zip code, drill down to characteristics such as families with young children, age range between 35 and 55, and income levels higher than $65,000. Make sure the target audience to which you are advertising matches your buyer persona. Targeting your ideal client with effective ads will generate the best quality leads with the least marketing budgets.

Choosing the Right Steps

Each one of your ads needs to include a concise response mechanism. It is a fancy way of saying you need to tell people what to do next and how to do it. Every single advertisement needs to be attractive and convincing to motivate candidates to take action. Move people to the next step. Persuade them to provide you with their email address, mobile phone number, or mailing address.

Consider the following ideas to generate responses and capture potential clients' information:

- The next step could be an invitation to your Unique Value Invitation (UVI) page (see chapter 3: DISCOVER YOUR CORE VALUES). Also, you can add a phone number on impression-ads to capture hot leads by asking respondents to call you or send you a text message. Use pay-per-click ads to drive people to one of your landing pages where they can request an educational free report. Grab their attention with compelling topics, such as *"Eleven Mistakes Most Sellers Make When Selling Their First Home." "Eleven Little-Known Tips for Buying Property Under Market Value (And When to Use Them)."* Be strategic and gain contact information and generate hot leads by choosing the right next steps.

- From any marketing material, you can use the next step to capture prospects' mobile phone numbers. Utilize call tracking technology to save people's phone numbers when they request information about properties you have promoted or featured on your social media. Call tracking allows you to display a toll-free number that individuals can call for pre-recorded information on a particular house. When they do, the technology grabs the caller's phone number and sends it to you within seconds so you can call them back. It's genuinely a brilliant innovation for lead generation. For more details, see the "Call Tracking" section under "Marketing Automation" in chapter 6: SHARPEN YOUR AXE.

Congratulations, now, everywhere people turn, there you are! The number one real estate agent in their area! The more they see you, the more credible you are, and the more they would want to hire you.

Success is not about how many people you know; it is about how many people know you, who you are, and what you do. Since you cannot physically call, text, and email everyone in your network regularly, you will let the digital world handle this for you by being omnipresent.

Remember, there's no secret formula for advertisement. Have fun! Execute and test different strategies, tactics, and campaigns. It is shocking to which ads people respond. Sometimes, you will make the most money from the least expected marketing campaign. Target diverse groups in multiple locations and test different content, images, and videos with various ads. Don't copy ads from agents in your area. What works for others may not work for you. Also, if your target audience resonates with a particular piece of information, it doesn't mean it will work for all people. Likewise, what seems viable in your area may not be relevant in other territories. To get the most out of your marketing dollars, you need to launch, test, measure, optimize, rinse, and repeat—quickly. Building a reliable lead generation system that works for you and your team begins with you taking actions and executing the right steps.

> *Take the Next Step*
> *Execute Tactic 9.10: Generate Leads through Paid Social Media Ads*

Generate Leads through Groups and Focused Communities

Pick a place where your potential buyers, sellers, customers, partners, and REALTORS® can collaborate, benefit, and enjoy your community. Focus on one niche (e.g., selling real estate) or one target audience (e.g., sellers). Then, expand to more niches, if, and when, your audience asks.

Provide your community with valuable content and educate them on your niche and services. Create a different theme for each day. For example, Monday could be about the selling process, while Tuesday might be about selling tips. Wednesday, you can share content about the closing process, and Friday can be a *Poll of the Day*. Be creative and try different themes to see what sticks.

Starting with a few fans and subscribers is normal. The most crucial step to build trust, respect, and rapport is to show up and be consistent. Show value over time to form a loyal tribe with a highly focused community around your brand. Developing a fan base will help you become the authority leader in your real estate marketplace.

> **Take the Next Step**
>
> Execute Tactic 9.11: Generate Leads through Groups and Focused Communities

Legal Notice

I am not an attorney neither do I provide legal advice. For your own sake, run all your marketing copy through your legal department. Staying legal is a must. The real estate industry is highly regulated. Being banned by the government or from advertising online will put you out of business. Do not take the law lightly. Do your due diligence and adhere to the Federal Fair Housing Laws, the U.S. Department of Housing and Urban Development (HUD) guidelines, and any other enforcing laws in your state. Most importantly, do not discriminate! After all, you're dealing with people just like you. Be ethical and honest. Be personable and treat others how you want them to treat you.

The Law of Attraction

Understanding the law of attraction is a fundamental necessity for your success. If you want to change your life, and empower yourself to create a fantastic future, then you need to understand your role in the law of attraction.

The law of attraction says you will attract into your life whatever you focus on. Whatever you give your energy and attention to will come back to you. Therefore, if you stay focused on the good and the positive, you will automatically attract better and positive things in your life. If you focus on lack and negativity, then that is what you will draw in your life.

When you apply this principle to your real estate business, your efforts, thoughts, and energy need to be in tune with what you want to attract. If you wish to target sellers in luxury neighborhoods, then you need to think, eat, and behave like those sellers. If you want to target buyers who want to buy multi-million dollar homes, you need to drive around those neighborhoods, learn about them and the surrounding attractions like you live there. Similarly, if you want to target an audience that is mainly lawyers, doctors, or engineers, then you must become part of their network.

You do not have to own a multi-million home to sell a luxury home. Owning a yacht doesn't have to be one of your goals to sell beach houses. You need not be a lawyer, a doctor, or an engineer to target such an audience. Instead, you need to learn about their needs and communicate what matters to them. Share such information, and watch your online traffic increase. You will then generate traffic of your ideal target audience. Once you see and live life through your clients' eyes, you will find innovative ways to attract them to your business and your life.

Traffic Types

You will be well on your way to generate high-quality leads and exponential success if you understand and master an essential concept called "traffic types." There are only three traffic types:

1. **OWNED MEDIA**: Traffic you own
2. **PAID MEDIA**: Traffic you control
3. **EARNED MEDIA**: Traffic you don't control

Once you understand how each traffic type works, you'll be able to direct the right traffic and the right customers to the right place. Then, you can leverage traffic sources to move customers to your website or mobile application where they can find viable solutions for their needs. By sending the right people to the right place, you would convert the highest number of visitors into customers.

A critical indicating factor why most online marketing strategies fail is because of not mastering the management of online traffic and their sources. However, you're now empowered. Your most important goal is to own all the online real estate activities and sessions within your territory. Owning web visits to your brand is how you will grow your database, sales, and eventually your business. The more online traffic you earn, the more leads you will produce, and the more deals you'll win.

Owned Media Traffic

Owned traffic is the best source to have and the most valuable for your business. That is when customers visit your office location, website, and mobile app. Owned traffic is the low hanging fruit to go after for growing your business. You don't have to buy that traffic from Google, neither Facebook. You can quickly email a group of connections within your database to generate instant traffic for your business. This book will help you master this significant step in building your real estate empire.

You have probably figured out by now that the more massive your database, the more money you'll make. That is why it is highly critical that you convert paid and earned traffic into owned traffic. Own your traffic and build an extensive contact list so you can lead your way to unstoppable success.

Although owned media takes some time to scale, it is cost-efficient and can build long-term relationships.

Here are a few examples of owned media traffic:

- Your website
- Your mobile application
- Your blog
- Your social media channels
- Your email lists

Paid Media Traffic

The next important traffic source is the one you control; in other words, the traffic your brand receives because of a paid advertisement. For example, yard signs, billboards, magazines—or online sources such as email ads, native ads, partners, affiliates, and joint ventures.

You control paid traffic because you can specify exactly where to direct it—for instance, a stranger clicking a paid social media ad and landing on your blog. However, you don't have ownership over that traffic—people browsing your social media profile. Nevertheless, you want to control it so you can own it by buying an ad-space and sending traffic to an owned media. Once a potential client gives you a ring or visits your online landing page, you then own that traffic.

While paid media could be costly and may have a declining response rate, if done accurately, it can help your business scale quickly.

Keep in mind; you must direct paid traffic to landing pages that have one goal; convert the traffic you control into traffic you own. Provide value and ask visitors to provide you with either an email address or a mobile phone number.

Here are a few examples of paid media traffic sources:

- Pay Per Click (PPC)
- Display Ads
- Retargeting Ads
- Paid Influencers
- Paid Content Promotion
- Paid Social Media Ads

Earned Media Traffic

The good news is you don't have to pay for traffic every time you need to own it. If you share a ton of content online, and it does well at serving your customers, you might hold a decent amount of traffic through earned media.

Consider random traffic that comes across your brand without knowing its source as earned media. Although, you should be tracking all traffic sources, including earned media from organic search, social media, and referrals. See chapter 6: SHARPEN YOUR AXE, section "Sharpen Online Tracking."

The best leads are usually referrals from happy customers that rave about you and your real estate services. Thus, earned media is the most credible and the most effective type of traffic in growing sales as customers become the best referral channel.

While earned media has its perks, it could also be very harmful if unhappy customers scold your brand. Make sure you keep your eyes on earned media and address any potential concerns immediately. If someone posts a bad review, reach out, respond, and ask what you can do to remedy the issue.

Here are some earned media traffic examples:

- Social shares, mentions, and reposts
- Reviews, testimonials, and word-of-mouth
- Publisher editorial
- Influencer outreach

Just like paid traffic, you want to take advantage of earned traffic and convert it to owned traffic. To do this, you need to direct your earned traffic to a centralized hub. It is the bridge between your landing page and the source of the earned traffic that connects people you don't know with your database. In other words, leverage earned media to convert visitors to subscribers. For example, if you post content on Facebook intending to direct people to your blog, it must have a lead form that captures your visitors' information. By using a lead magnet that will direct them to a landing page, and obtain their contact information. Notwithstanding, you must always convert earned traffic to owned traffic by getting people to opt-in.

Generate Leads Using Combined Traffic Types or Sources

Anyone can receive paid traffic and online leads by investing in Google, Facebook, or Zillow. Nonetheless, a mix of two or more traffic types can have a notable boost in more traffic to your brand. Therefore, combine earned with owned media sources to increase the number of leads collected consistently.

Furthermore, a better way to increase brand awareness and lead acquisition without having to dig into your savings is through dedication and persistent promotion. Here's an example that will exponentially increase your business. Suppose you wrote a relevant article you strongly believe buyers and sellers would love to read. You contact Inman, Zillow or Trulia and you pitch your piece. One of these giant leaders agrees to feature your article on their website. Congratulations, as an author with a credible publisher, your business has just earned a free advertisement. Thus, you've made a media relation. Now let's say your article has links directing readers back to your blog. When your publication receives tons of eyeballs and many social shares, your blog will receive referral traffic. Hence, you earn that traffic once readers are on your blog. Also, you will own a portion of those people who share their email address or phone number. Therefore, in this case, you have a combination of both earned and owned traffic types. Winning two traffic sources is a big game-changer at an affordable price. Write a valuable column, get it featured, and drive traffic back to your blog or centralized hub! Try it and witness lead generation at its best with combined traffic sources.

Take the Next Step

Execute Tactic 9.12: Generate Leads Using Combined Traffic Types or Sources

Generate Leads Using a Specific Traffic Source

Never rely on just one principal source for generating leads. You need multiple traffic sources for your lead generation strategies. Determine the types of audience you would want to farm, and figure out the best sources for those leads and the best communication channels to convert them to customers:

- **Type of Audience to Farm:** What sort of leads are you looking to generate? Pick only one persona for each strategy. For example, Buyers, Sellers, FSBO, Expireds, canceled, and withdrawn listings.

- **Source of Leads:** Which sources will you use to generate leads? Pick only one source for each strategy—for example, Content, Social Ads, Beacons, Live Chat, BoomTown, Zillow, Realtor.com, Chime, Cole Realty Resource, Direct Response, or Geo Farm.

- **Marketing/Communication Channels:** Which marketing channel will you use to communicate with people and generate quality leads? You can deploy several channels for each strategy, but you must track each channel separately to determine which one caused the most deals—for example, Dialer, Voicemail Drop, push notification, Text Messaging, Email, Facebook Messenger, Live Chat, and Mobile app.

Test each traffic source separately to determine which traffic type performed best in winning customers—for instance, which traffic converted better and brought the highest number of customers?

- Tactic 9.12 using the "Letter to the Editor (LTTE)" Strategy
- Tactic 9.13 using the "Cole Realty Resource" Strategy
- Leads you bought from Zillow, Trulia, or Realtor.com
- Leads from social media

Take the Next Step

Execute Tactic 9.13: Generate Leads Using a Specific Traffic Source

You can follow the same strategy with other sources. For example, use the 6,000 contacts as a targeted list source. Attract potential buyers to listings in your area, driving the traffic to landing pages using voicemail drops, email, text messages, or paid ad campaigns. You can execute the same strategy attracting sellers to pre-qualified buyers looking to buy in your area.

With the right audience, traffic sources, and marketing channels, you can take advantage of all tactics and lead generation strategies that I shared with you. The possibilities are limitless to generate quality leads in mass.

CHAPTER TEN

Nurture Meaningful Relationships

"Magic happens when you don't give up, even though you want to. The universe always falls in love with a stubborn heart."
—JM Storm

In chapter 8: PLANT MORE SEEDS, you reached out and introduced yourself to thousands of people. You created brand awareness through demand generation. In chapter 9: GENERATE HIGH-QUALITY LEADS, you attracted attention and generated quality leads through marketing and advertising. In this chapter, you will create desire by nurturing your leads through lead activation and engagement.

Lead Activation

How do people derive value from coming across your brand before becoming customers? How do you get them to experience the first aha moment—the first taste of value—so they can consider hiring you as their real estate agent?

Most real estate companies are so concerned about lead generation; they pay little attention to the activation process. What few realize is that activation plays a significant role in customer acquisition.

So, what is lead activation? Lead activation is the bridge that connects lead generation with customer acquisition (the moment a lead signs your listing agreement and becomes a customer). It is the nurturing process of generating active engagements with cold and warm leads before they reach the point of the buying intent—signing the contract—in which they become customers. The lead activation process purposefully advances engagements and increases their probability of becoming high-quality sales opportunities.

You realize leads are in the activation process when they experience great impressions with your brand. Also, activation happens at the point in which prospects realize the real values from interacting with you or your real estate company. Lead activation shifts people from the marketing department to the sales department through cultivation. Nurturing your database moves contacts through the marketing funnel into the sales funnel. You then activate those who seem like viable opportunities. Consequently, Marketing Qualified Leads (MQLs) become Sales Qualified Leads (SQLs). The more you boost the activation number, the more your sales channel becomes profitable.

Optimize your activation process similar to how you would improve your advertisements or any of your marketing efforts. Simplify the onboarding process and the user experience with your brand— successfully onboarding as many people as possible by transitioning them from marketing to sales. When you nurture your database and move people from cold leads to sales opportunities, you are progressively activating them to become customers.

Value Triggers

Helping your customers experience the full value of your real estate services is critical in ensuring their loyalty. Increase the number of clients and extend their lifetime value by first uncovering your engagement steps, and second foreseeing the value triggers.

A Customer Lifetime Value (CLV or often CLTV) is a prediction of the net profit attributed to the entire future relationship with a customer. A value trigger is a unique offer that triggers emotional acknowledgment and realization of a benefit your audience did not expect. For example, the aha moment that Sally the Seller didn't expect when she realized the value of the Cancellation Guarantee Policy. Rose the Realtor mentioned the policy to overcome Sally's objection of signing the listing agreement. Since Sally wasn't aware of the plan, she experienced an aha moment—a value trigger that made Rose more credible.

Uncover engagement steps through various levels of the marketing and sales process—whether in person or online. For example, create a custom Checklist of Interest that Bob the Buyer can complete to find him the best home. Hand-deliver a real estate selling guide that educates Sally, the Seller how to get the highest price for her house! Which benefits are you using to that will demonstrate your values and entice your clients into taking action? What messages you use that will trigger their curiosity are highly crucial in increasing engagement and getting them closer to becoming your clients.

How are you educating your potential sellers on the series of action steps they have to take from being interested in selling to the listing presentation, and then to close? Where in the nurturing process do you expect sellers to experience their aha moments?

How are you engaging with your potential buyers from the moment they search for their dream home to signing the contract? What are the actions that these potential customers take, and along the journey, which values do you anticipate triggering their aha moments?

Those aha moments are your value triggers! The more your audience senses them, the more they'll engage with your brand. The more they'll see value in your product, the closer they are to the magic moment.

To optimize the activation process and win more potential sales opportunities, you must create more aha moments and keep the magical moment top of mind.

The magic moment is the most significant value trigger! It is what your customers will experience when you accomplish their ultimate goal. For example, for Bob the Buyer, it is when you hand over the keys to his new home. For Sally the Seller, it is when she gets paid after you sell her house. How about Rose the Realtor? Her magic moment as a real estate agent is when she gets paid after closing the deal.

Create more value triggers to get customers to experience more aha moments as quickly as possible. The faster they reach their magic moment, the sooner you will experience yours.

> *Take the Next Step*
>
> *Execute Tactic 10.0: Identify Your Value Triggers and Create More 'Aha!' Moments*

Activation Metric

Also known as the North Star Metric (NSM). The Activation Metric is the leading metric that best captures the core benefit your product offers your clients. Optimizing your marketing efforts to grow your NSM is crucial to driving sustainable growth across your full customer base. The most indispensable sales-metric you must define and measure is your activation metric. For example, the number of potential customers who have signed the listing agreement! Your activation metric would be the number of listings or potential sellers and buyers you have in your pipeline.

Once you define and measure your activation metric, you and your teams will need to optimize the inputs and the steps of the onboarding process that lead up to activation quickly. The more precise your definition is, the easier it will be to prioritize marketing and sales campaigns to increase it. Consequently, activation becomes a positive experience of adding sales opportunities to your pipeline! That said, how do you activate more leads?

Before you start a new onboarding campaign, ask yourself:

- How does this onboarding campaign increase my/our activation metric?
- Will this campaign increase my/our activation metric more than other campaigns?

Articulating marketing objectives to reach the activation metric will provide huge dividends to you and your real estate company. Laser-focus your activation efforts on encouraging customers to take the next step! Get them deeply involved in the process so they can feel that they are in control and a big part of the adventure.

For example, "*Sally the Seller, the next step is to sign the listing agreement, would you like to do so in person or digitally online?*" You activate a quality seller lead when Sally signs the contract regardless of whether in person or online. Asking this question by phone, in an email, a text message, or on social media is the action that will get you closer to a higher activation metric. Also, it is one of the steps that get your clients closer to experiencing their magic moment. Making decision-based closing questions a necessary level of the onboarding process helps you reach more activations and faster closings.

By combining your customers' aha moments to reach their magic moments, you will convert more leads to active customers. Stay focused on your activation metric and think of creative ways to activate more clients in less time!

Take the Next Step
Execute Tactic 10.1: Activate More Customers with Value Triggers

"*Your perseverance demolishes all obstacles and breaks all barriers.*"
—James Tyler

The "Flywheel" Concept

Thanks to Jim Collins for the "Flywheel" concept developed in his book "Good to Great." It elaborates on the importance of momentum and the necessity of gradually moving leads through the closing journey until they become raving clients.

Picture a huge, heavy flywheel—a massive metal disk mounted horizontally on an axle, about 30 feet in diameter, 2 feet thick, and weighing about 5,000 pounds. Now imagine that your task is to get the flywheel rotating on the axle as fast and as long as possible. Pushing with great effort, you get the flywheel to inch forward, moving almost imperceptibly at first. You keep pushing and, after two or three hours of persistent effort, you get the flywheel to complete one entire turn. You keep pushing, and the flywheel begins to move a bit faster, and with continued great effort, you move it around a second rotation. You keep pushing in a consistent direction. Three turns ... four ... five ... six ... the flywheel builds up speed ... seven ... eight ... you keep pushing ... nine ... ten ... it builds momentum ... eleven ... twelve ... moving faster with each turn ... twenty ... thirty ... fifty ... a hundred.

Then, at some point—breakthrough! The momentum of the thing kicks-in in your favor, hurling the flywheel forward, turn after turn ... whoosh! ... its own heavyweight working for you. You're pushing no harder than during the first rotation, but the flywheel goes faster and faster. Each turn of the flywheel builds upon work done earlier, compounding your investment of effort. A thousand times faster, then ten thousand, then a hundred thousand. The huge heavy disk flies forward, with almost unstoppable momentum.

Now suppose someone came along and asked, "What was the one big push that caused this thing to go so fast?" You wouldn't be able to answer; it's just a nonsensical question. Was it the first push? The second? The fifth? The hundredth? No! It was all of them added together in an overall accumulation of effort applied in a consistent direction. Some pushes may have been bigger than others, but any single heave—no matter how large—reflects a small fraction of the entire cumulative effect upon the flywheel.

The merit of the story is that it takes persistence, a tremendous amount of effort, and an immense amount of time to get the momentum going. In a nutshell, there is no such thing as one massive push to breakthrough.

That said, generating leads and expecting to close them with one sales strike is a myth. Just how ineffective would lead generation be without first creating demand and awareness? The sales numbers will plummet without early nurturing your database. Think of Jim's Flywheel concept and imagine how many times you must push that thing before the breakthrough point. Likewise, you must cultivate relationships consistently until you activate each contact in your database, turning them into quality leads and loyal clients. Hence, the need for lead nurturing!

Lead Nurturing

According to Forrester Research, companies that excel at lead nurturing generate 50% more sales-ready leads at 33% lower cost. Valuable content, how well, and how fast you respond to leads are the most important factors to remember in the client nurturing process. To convert people to customers, you must measure their interest and their buying intent.

While you can call and chat with people about your real estate services, online content is one of the best methods used to measure intent. Online content can easily be trackable, providing useful insights into potential customer behaviors. Deliver valuable content to your database that will advance them a step further within the conversion funnel, the marketing, and the sales process.

Deliver such content via any of the offline and online communication channels shared with you in previous chapters. Nurturing your database doesn't mean only to send monthly email newsletters. Be omnipresent! Send personalized direct mail to people who intend to sell or buy within three months. Also, automate the nurturing process via social media (messenger, chat, and direct messaging), push notifications (web and mobile), drip campaigns (email and SMS text messaging), and voice campaigns (phone dialer and voicemail drops).

To ease the nurturing process, it would make sense for you to group your audience in your database into different categories—for example, sellers, buyers, FSBOs, and Expired Listings. You will also need to segment them based on their needs and how soon they are looking to satisfy them. A "need" could be an interest in selling in less than three months, within six months, a year, or over a year. Segment your contacts so you can deliver personalized content to each specific group. Splitting your network into different groups will also make the communication and the follow-up process much more convenient for you. Your potential clients will see how relevant you are in your correspondence. They will resonate with your content and therefore engage more.

By segmenting your contacts and engaging with your leads, you will have a great sense of how often you should follow up with each group. There is no right and wrong in how often you must check in with your network. You are dealing with people—not robots. So, it all depends on your audience, their needs, and your efforts. For example, communicate and follow up with "hot" leads who committed to buy or sell in the next three months, touching them at least once a week via every communication channel: mail, email, text, push notifications, voice, phone, and social.

It would make more sense to follow up with the "warm" group of leads who are buying or selling within six months. Touch them at least once every two weeks via email, text, voice, and social. Otherwise, schedule everyone else on a monthly drip campaign. Get in touch with them once a month via email, social, and perhaps SMS text campaigns.

How to Start the Lead Nurturing Process?

Taking you back to the "Introduce Yourself" section in chapter 8: PLANT YOUR SEEDS, you did a direct outreach with your "Soft Introduction" email to 6,000 people in your database to create demand. Some probably answered, and most didn't. After your virtual assistant reached out to your entire database, he or she tagged your contacts with the proper status. For example, cold leads, warm leads, hot leads, past customers, family, friends, and partners. See chapter 8: PLANT MORE SEEDS, Section: Your First Touch.

Now, you will create a nurturing program for each group in your database. Let's say you will develop a semi-annual drip campaign that will extend over the next six months. If you're nurturing cold leads and sending one communication per month, you'll need six templates to strengthen your relationship with that group. Similarly, you'll need 12 drafts to cultivate your relationship with those who are buying or selling in the next six months, sending them one communication every two weeks. Finally, your most targeted nurturing campaigns will need to focus on the hot leads. You'll also need 12 templates to deliver over to this group who is buying or selling in the next three months, touching them once a week.

If you think about this nurturing process thoroughly, you'll notice that you only really need to create 12 templates. You can write those 12 templates in email formats, then repurpose them for SMS, push notifications, and then direct messages on social media.

Tag your leads with cold, warm, and hot so you can have a starting point for the nurturing process. For example, nurture cold leads with email template #6, warm and hot leads with email template #1. Therefore, to run effective lead nurturing programs, you must leverage marketing automation by delivering valuable content to serve your audience's needs over the course of their anticipated lifecycle and beyond.

> **Take the Next Step**
> *Execute Tactic 10.2: Create Lead Nurturing Email Templates*

How Do You Set up Your Marketing Automation Program?

If you haven't signed up for one of the marketing automation programs recommended in chapter 6: SHARPEN YOUR AXE, you will need to do so, pronto. If you want your real estate business to scale, you've got to automate most of your marketing efforts. Nowadays, the marketing automation platform is a must-have for every business.

The Marketing and Sales Funnel

You can't possibly automate the lead nurturing process without first building your marketing and sales funnels. A funnel is nothing more than a conversion process your customers go through from the beginning to the end — your initial contact, to becoming customers. It is the journey your customers experience during the life cycle of the entire real estate transaction.

The marketing and sales funnel comprises multiple layers. The number of layers varies depending on each specific industry and business model. While some businesses may require five or six layers, fundamentally, there are only three layers to the funnel: top, middle, and bottom.

Strangers reside in the top layer of the funnel. You would introduce yourself to this group of people to build awareness and move them to the middle layer. The middle layer comprises people who know of you and what you do. Some of these people are cold leads; others are warm. You would categorize warm leads as Marketing Qualified Leads (MQLs). Once MQLs are ready to commit, they become hot leads. You would categorize hot leads as Sales Qualified Leads (SQLs). The transitioning of warm leads into hot leads moves contacts from the middle to the bottom layer of the funnel. That is when the marketing department hands over MQLs to sales staff to turn them into customers. If they don't become customers, you either kick them back to the middle layer for further nurturing, or burst them out of the funnel as bad deals.

The goal of the top of the funnel (ToFu) is to attract your most relevant audience. Educational content is the most common form of material used to attract prospects to the ToFu. Blog articles, eBooks, or cheat sheets will do wonders in attracting early-stage home buyers and sellers. Strangers enter the funnel once they exchange their contact information for the valuable content you provide.

On the other hand, the middle of the funnel (MoFu) has two functions: Qualify your leads and build your credibility. In this stage, you want to encourage your audience to provide information about their needs. It is critical that you understand their pain points so you can position yourself and your company as the solution to their problems. Nurturing campaigns, white papers, quizzes, sold listings, and videos are great for progressing leads from cold to warm. Use past case studies, customer reviews, and testimonials to build trust and rapport.

Furthermore, warm and engaged leads who qualify for your real estate services are a top priority. Move those people from the MoFu to the bottom of the funnel (BoFu). In this layer, you must have one-on-one engagements with your leads. Personalize your communications to qualify your most interested leads. The goal of the BoFu is to convert hot opportunities to customers.

Now that you understand the funnel; next, you will learn how to use it. Business consultants created the marketing and sales funnel to help marketing and sales track the progress of their leads through the buying life cycle. As a real estate agent, all you need to find out is whether a prospect is ready to buy or sell in 12 months, six months, or three months. You can dig deeper and get technical if you want to track people who are committing in over 12 months or fewer than 60 days. I recommend you keep the tracking process simple. Categorize the people who are engaging in 6-12 months as "cold" leads, 3-6 month as "warm," and 0-3 months as "hot" leads, aligning these three groups into the Top of Funnel (ToFu), Middle of Funnel (MoFu), and Bottom of Funnel (BoFu) respectively.

Now that you know how to use the funnel and its layers, how do you automate it? Previously, you've written your 12 content templates. Since email is the most commonly used channel for nurturing leads, the same rules apply for SMS, push, and social media in case you decide to leverage other cultivation channels. The first step to automation is to add those 12 templates to your marketing automation program.

Most businesses want to create flashy email templates with images, animated illustrations, colored fonts, and pretty buttons. They invest in design, coding, and HTML implementation only to find out that most emails are landing in the SPAM folder. DO NOT fall for this trap. People like short and simple messages. Google and other well-known email service providers prefer text emails. The more succinct and straightforward your email template, the better! Text-only version emails work best. Type your emails on a notepad and paste them into your marketing automation platform. That is it—no design, coding, formatting, complex layout requirements, or responsiveness for mobile is needed.

Here's what's important! Each template must have a call to action—this could be a link. Once clicked, you can determine which email template to deliver next. Suppose you sent the hot leads email template #1, those who open and click the link in that email will then receive email template #2 in 3 days. If they open and click the link in the second email, they will then receive email template #3 in 2 days and so on. For those who open and don't click, send a different email to see what resonates with them. For those who don't even open your email, you would need to send the same email with different subject lines to see what gets them to open your email. Test different subject lines and different content to see what gets you the most conversions.

> **Take the Next Step**
> *Execute Tactic 10.3: Set up Your Marketing Automation Program*

Testing with Automation

All Marketing Automation Platforms (MAPs) come equipped with drip campaign capabilities where you schedule the entire six months of email, text, and push campaigns in advance. MAPs also provide you with testing features such as A/B Split Testing and Multivariate Testing (MVT). Depending on the MAP you are using, check to see which testing features are available to you, and enable them within your campaigns.

Without getting too technical with testing, the goal is to gain more activations, advancing your leads to the next step in the real estate process. You are sending nurturing messages, so contacts within your database can move through the real estate buying or selling lifecycle. If they don't open your email, change the subject line until people take measures. If they open the email but don't act, improve the body content and the calls to action so you can entice them into clicking or calling. Applying A/B and multivariate testing to your marketing automation campaigns is nothing but changing your message elements around to see what works with your audience moving them through the three stages of the funnel to close.

Lead Grading and Scoring

There are two other topics you may hear about in marketing automation: lead grading and lead scoring.

Implement lead grading into the automation process to determine the quality of leads you are nurturing. Calculate lead grades based on people's demographics and qualifications.

Suppose you're licensed to do real estate in California. You receive an online lead that is interested in buying a home in Arizona. Here, you set a rule in the marketing automation platform to score out-of-state prospects to zero since you hold a California license only.

MAPs come with numerous data inputs where you can fully customize the lead grading system. You have the convenience of setting your qualifying and grading rules. For example, if you set the maximum grade to 100, you can automatically subtract 100 from all leads who aren't buying or selling in your state. Otherwise, you can add whatever scores you want based on qualification factors. For instance, based on the home price or value, whether the lead is a buyer or a seller, the time frame to commit, geolocation or territory, and whether the client is qualified with a lender.

Here are a couple more examples to elaborate on grading your leads: You can assign a grade of 10 for homes lower than $500K, a grade of 15 for homes between $500K and $1M, and 20 when the home value is over $1M. Similarly, you can assign a grade of 10 for cold leads, 15 for warm leads, and 20 for hot leads ready to take action in less than three months. Give 5 points to buyers and 15 to sellers. A grade of 15 if the contact is within your farm, and 5 otherwise. Add 20 points if the lead is a referral, or 5 otherwise. Repeat the process as needed to grade your database with the sum points of 100.

Since referrals don't cost as much as leads sourced from other advertising channels, a perfect lead would be a referral (20). Add (15) if the referral is a seller, that is selling in your territory or farm (15), in less than 3 months (20), and their home value is over $1M (20). Then, in this case, the total grade would be 90 points (20 + 15 + 15 + 20 + 20). This prospect is someone you would want to call immediately. However, how would you know to call right away? Set up a rule in your marketing automation platform to notify you via text message every time you have a lead with a grade of over 50 points. Powerful!

Further, to score contacts within your database, instead of using lead quality and demographics, you calculate the lead score based on behaviors. The more an individual engages with your brand, the higher the rating. You can split your cold, warm, and hot lead scores to 20, 50, and 80 points consecutively.

Assign score points based on leads interactions. For example, 1 point for each website visit, 3 points for each email open, 5 points for each link clicked in an email, and 7 points for each link-click in an SMS message or push notification. Add 10 points for each eBook downloaded of a landing page, 20 points for each appointment booked, and 25 points for meeting face-to-face. Again, set up automated alerts to follow up with leads that have scored higher than 50 points.

Lead grading and scoring not only will help you nurture your database with smart and automated messages but also remind you to follow up with those who matter most.

> **Take the Next Step**
>
> *Execute Tactic 10.4: Move People through the Marketing and Sales Funnel*

There are many moving parts and requirements in the above tactic. From locating your 6,000 cold leads, their email addresses, mobile phone numbers, and social media profiles, to introducing yourself, sending them the first touch, and automating the entire marketing strategy.

From communicating via email, social media, text messages, and push notifications to tracking user behavior via email, landing pages, and social media; every single step is a highly crucial part of the entire domination strategy.

Mess up a stage, and the whole approach may fail! Once you *test*, *measure*, and *optimize* each level of the plan, the entire strategy becomes more reliable and robust. The more you execute these *three* steps, the more conversions you'll have, and the more clients you'll earn. Perfect your nurturing funnel and watch your real estate business skyrocket.

When creating your marketing and sales funnels, do not limit yourself to just emails. Put your leads through various automated drip campaigns. Take advantage of email and text messaging, web and mobile push notifications, and private online messaging like chat and chatbots.

Remember to automate the ToFu and the MoFu layers. However, limit your automation process within the BoFu layer, especially sales and customer service. Don't automate the process of converting hot leads to customers or when dealing with new customers. To build effective long-term relationships, you must always personalize your real estate services to maximize your phone and face-to-face communications.

What Should I Say to People?

A simple question, yet asked by so many real estate professionals. You should deliver relevant content that resonates with your target audience. Once you speak to their needs, you will earn their trust and their business.

If your target audience is sellers, mentally walk through the selling process. Write *educational* guides and blog articles about each step of the selling process to educate your audience. For example, what to look for when hiring a real estate agent, how sellers get the highest price for their listing, and the things they need to know to stage their house correctly. Also, when and how do sellers get paid, and how to sell and buy another house, concurrently! The complete home selling guide should also cover reasons why hiring you as an agent is far more advantageous than a For Sale By Owner (FSBO). Don't forget to include your contact information so potential sellers can call you when ready! Turn these topics to blog articles, social media posts, downloadable PDF documents, and videos to capture more traffic and leads. Also, every time you write a publication, turn it into a podcast or a video. This way, you can nurture your database via email, text, and social media. Follow the same concept if you're a selling agent so you can educate your potential buyers on homeownership and the various home buying topics.

Educational content helps not only your audience but makes your life more comfortable as a real estate professional. Educated customers will save you time and hassle during the process. When you become a busy agent, you won't be able to spend adequate time on the phone, educating potential leads. With content, you will have the advantage of directing those people to your blog. This way, they can learn about the selling or buying process without you having to get involved. Educational content can also help you share information your audience needs via email and social media. Write and educate your audience with content. Valuable content will positively impact your real estate career, guaranteed!

Another form of content would be *transactional*. It is the form of content you use to interact with your network regarding transactions currently in process. Interactional content is the most effective personalized content and the easiest to create. Compose them once and then automate them. For instance, suppose you have just received a phone call from a brand-new potential seller. The call went well! You get off the phone, and you compose a *"Thanks for reaching out"* email appreciating the caller's time. Imagine you just met with a potential buyer. Again, shortly after the meeting, you email the prospect, *"It was great seeing you,"* thanking the person for their time. You have just met a client with the home inspector, or the appraiser. One more time, you send an email to parties involved thanking them for their time. Before the appraisal or an inspection, you email the potential buyer, *"What to expect during a home inspection?"* Do you see a trend here? These transactional emails are the same throughout every single transaction. At most, you'll have minor changes, yet, most real estate agents copy and paste emails repeatedly. Why not create email templates and automate the entire real estate transaction from A to Z? Save yourself the time of writing your transactional communications all the time. Search Google for "canned pre-written real estate emails." Navigate through a plethora of resources, re-write relevant templates, and automate your communication with your clients.

Besides educational and transactional content, you may share information that resonates with professionals such as partners, vendors, and affiliates. By doing so, you will build rapport and gain their trust. Stay top of mind by sharing news and trending information related to their specific space. For example, sign up for Google Alerts to receive relevant news you can share with your favorite lenders, loan officers, appraisers, title representatives, escrow officers, inspectors, and contractors.

Find information that is worth sharing with professionals in your database at least once a month. Write a blog article about it and share it via email, text, and social media. You can also influence professionals by sending them referrals. Another tactic is to have weekly or monthly lunches or coffee meetings with two different individuals that don't know each other. Becoming a liaison introducing and connecting people is a powerful way to connect. A simple method to grow your network, yet, few real estate agents execute on it.

It is imperative to touch every single contact in your database at least once a month. There should never be a case where 30 days pass, and you don't reach every person. Communicate with potential customers, existing clients, partners, friends, and family members regularly. Keep them up to date with market stats, trends, news, and upcoming innovations relevant to your real estate industry. Be a useful source that every person within your network looks forward to receiving your text messages and emails. The least you can do is reach out and say, *"Hey, just checking in to see how you're doing!"*

The next time you are about to send an email, a text message, or share a post on social media, ask yourself, *"Will I care to read or watch this?"* If the answer is *"No,"* rethink your content and your strategy. People are more likely to tune to the "what's in it for me?" or WIIFM channel than self-promotional content. Having a Facebook wall and an Instagram feed filled with random real estate *memes* or inspirational quotes makes you a mediocre real estate agent. While it is acceptable to have an occasional open house and featured-listing posts, people want to consume valuable information that benefits them. Keep in mind that real estate is just like any business. It isn't about you! It is about your customers, their needs, and wants. Once you put your customers first and remove self-interest from your business model, you will build meaningful relationships and eventually succeed.

For more information on how to write valuable and enticing content, please refer back to chapter 5: SPEAK YOUR MESSAGE WITH INFLUENTIAL CONTENT.

By now, you should have moved most contacts from the Out-Of-Network to the Network spreadsheet, or your CRM. Meaning, you should have driven people from awareness to the discovery stage. Next, you will use marketing automation to nurture those leads and turn as many of them into clients. Therefore, moving people from your Network to your In-Network! It is where your family, friends, acquaintances, colleagues, clients, and partners reside.

Take the Next Step
Execute Tactic 10.5: Nurture Your Database with Valuable Content

Follow strategies and tactics shared with you in these pages to turn cold leads into warm leads, and warm leads into hot leads. In case you're curious, in the next chapter, you'll turn hot leads into customers. However, before we get into sales, let's talk about remarketing.

Remarketing

Remarketing (also known as retargeting) is the tactic of serving online users paid targeted ads based on their historical online activities and behaviors.

To serve people with relevant remarketing ads, a savvy web developer must install a script on your websites. Programmers technically refer to it as Javascript tag—commonly known as a pixel and formally as an insight tag.

Large digital advertising networks such as Google, Facebook, AppNexus, or OpenX will place a cookie on the user's browser. A cookie is a small file that comprises a few lines of Javascript code—also, called pixel. Advertisers save them on users' computers and devices with data about their browsing activities and online behaviors. The cookie, or pixel, informs the advertising provider to serve specific display ads through an ad exchange.

An ad exchange is a technology platform that facilitates the buying and selling of online media advertising inventory from multiple ad networks. It delivers purchased ads to people based on the web pages they've visited, and online products with which they've engaged.

The paid ad agencies I recommend using for your remarketing campaigns are Google (including Ads, Display Network, DoubleClick, and YouTube), Facebook Audience Network (including Facebook and Instagram), LinkedIn, Outbrain, and PerfectAudience.

Here's a quick overview of remarketing in action:

1. Users visit your website, blog, or landing pages.
2. They leave your websites without completing a single call to action.
3. They browse other sites.
4. Your ads appear on other web pages users visit online or mobile devices.
5. Having a previous interest in your real estate services, some of these users will click on your ads and return to your website, blog, or landing pages.
6. You will have a second chance at generating leads.

"Retargeting is a tool designed to help companies reach 98% of users who don't convert right away. After retargeting with an ad, website visitors are 70% more likely to convert."

—retargeter.com

The fact that there is a 70% chance to convert is because remarketing will allow you to segment people who visit your website in various ways. You can create specific ads to target people who have visited individual pages or retarget by custom audiences, time spent on the site, and the number of visits. For instance, segment by page to retarget people who have searched your IDX solution with a specific "city" in the Universal Resource Locator (URL) of your website. Segment by email as you can embed a pixel into HTML emails and retarget people who have opened specific emails. Distinguish and segment people who have clicked distinct buttons on your websites or searched for a particular word.

Marketers call the search term that helps you track user behaviors the "QueryString." Google implemented this tracking mechanism in what they call the Urchin Tracking Module or (UTM) parameters.

"UTM parameters are five variants of URL parameters used by marketers to track the effectiveness of online marketing campaigns across traffic sources and publishing media."
—Wikipedia

Another fact that proves remarketing is so powerful by giving you the ability to reach the 98% of people who didn't initially convert on your sites. According to Similar Tech, DoubleClick is unquestionably leading in Top 10K Sites, Top 100K Sites, Top 1M Sites, and The Entire Web. DoubleClick has over 2.3 Billion websites on which they can display your ads. Chances are, Google DoubleClick, also known as "Google Marketing Platform," is tracking every single user who visits your site—and so are YouTube, Facebook, Instagram, and many others. Yes, that is a good thing, and that is what you'll leverage.

When advertising on Google, Outbrain, or PerfectAudience, you must ignore irrelevant websites within their networks. For example, advertising on news and real estate related sites would be more relevant than gaming or apparel websites. Excluding specific segments such as people outside your geolocation can also save you a ton of money on your advertising budget. Leverage this exclusion feature to ignore particular publications and exclude irrelevant audience. Advertise to people at a granular level and earn higher returns on your remarketing investments.

Why Is Remarketing Important for My Real Estate Success?

Remarketing delivers increasingly high-relevant and targeted ads. Retargeting campaigns are very effective at promoting a specific product, offering, or brand. Retargeted ads remind not only online visitors of their desire to engage with your real estate brand but can facilitate the transition to the next step of the conversion funnel.

If users were searching on your IDX solution page and leave your website without getting in touch, in the absence of remarketing, you'd lose those potential leads forever. A remarketing campaign can present advertisements to them immediately anywhere on the web. Retargeting ads can lead those users back to your landing pages. Then you will have another opportunity to incentivize them and move to the next stage of the funnel. For example, entice them to download a Comparative Market Analysis (CMA) report in exchange for their email address.

Sync your customer segments with ad networks like Google and Facebook and the likes so you can extend your media ad campaigns with automated lookalike audiences. All advertising agencies offer similar syncing capabilities where you can deliver highly customized ads to segmented audiences such as sellers, buyers, and FSBOs. You can even personalize your ads based on segmented contacts that are cold, versus warm and hot leads.

Take the Next Step

Execute Tactic 10.6: Set up Retargeting Ads on Facebook

Execute Tactic 10.7: Set up Retargeting Ads on Google Ads

CHAPTER ELEVEN

CREATE PROMISING OPPORTUNITIES

"Formula for success: rise early, work hard, strike oil."
—J. Paul Getty

Everything until this chapter was about business development and marketing. Nothing we covered this far was apropos of sales. Doing everything right until this moment was half of the battle, to dominate, you must master the other half—serving clients well through the real estate sales process and beyond.

Lead Conversion

The most beautiful summation of the top-performing real estate agents is their focused mindset on making a difference each day, again and again, steadily. They all have the right mindset to attain their SMART² goals. The habit of continued-performance and consistency in their daily actions is a natural regime rather than a real estate career.

Most successful agents do not leave the office until they set up an appointment with a potential lead each day! There is a trend at leading real estate offices where agents gather each morning before 8 AM to make hundreds of outbound prospecting calls. Others set weekly goals to finish up each term with a minimum of three listing appointments. Whichever goals you set for yourself, and whatever path you follow in advancing your real estate career, closing deals and getting big rewards is nothing but a result of consistently achieving those daily micro-commitments with the sole intention to serve your clients, not to sell them.

How do you reach breakthrough pushing that flywheel daily? By increasing your activation metrics!

1. Set up more interviews
2. Execute more contracts
3. Close more transactions.

You can improve your activation metrics when you take people offline and get them to say, YES. Did you ever have someone visit your website and instantly hire you? Agents only wish it was that simple, right? Closing a real estate transaction isn't easy. There's always a multi-step process to convince Bob the Buyer and Sally the Seller to sign that contract. Executing an agreement isn't enough. To monetize, Rose the Realtor must close the deal. Sometimes the process is pleasant, other moments, it is painful. One thing Rose knows for sure; Bob and Sally won't execute that agreement until they fill up her email inbox with questions, get on the phone a few times, and finally agree to meet in person after persistent follow up by Rose.

The more face-to-face consultations you have, the more signed contracts you will eventually win, and the more money you will make. Therefore, the lead conversion process comprises two steps:

1. Scheduling more face-to-face appointments
2. Getting more applicants to sign the contract

We will discuss the third activation metric in chapter 12: SEAL THE DEAL, under the lead monetization section. If getting paid is your magic moment, then *closing the deal* is the last activation metric within the process. Hence, the formula to your real estate success is simple; focus on your activation metrics of booking in-person interviews, getting more clients to sign, and closing more transactions.

> *"Selling is a dialogue; not a monologue! Listen, understand, and then serve."*
> —James Tyler

Prospecting

Often agents view prospecting as a lead generation effort when, in fact, it is a conversion tactic. To prospect, you must have leads. You prospect by calling them to schedule appointments. Therefore, converting leads to opportunities. There's your first activation metric!

What separates the top-performing, most successful real estate salespeople from everyone else is that they are out-selling and out-earning many times over, daily! They all share the same sales approaches and common skill sets. Their approach to revenue is to win much more sales consistently through prospecting.

Even when their pipeline is full, they never stop prospecting. Pipeline activity does not equal deals, and you never know what the future holds. Not every appointment turns into a signed contract, and not every signed contract turns into a sale. Therefore, prospecting is a must for all real estate professionals who want to win and dominate. If you're not prospecting, you're doing yourself a disservice.

Get Prospects on the Phone

To prospect more people effectively, and get far more potential clients on the phone, you must dial. Keep a dialog going, raise people's interest and curiosity, then share valuable information that will keep them engaged and finally convince them to meet with you.

Dial that number and make your daily phone prospecting process a fun routine that leads to relationships. Master the phone prospecting tactics that most successful real estate agents have in common.

Believe in Yourself

Are you feeling intimidated? Face your fears, head-on. Be confident and pick up the phone. Tackle the massive amount of rejections by considering prospecting a challenging game you play every day. Play the big game to score big wins. Take risks and don't be afraid to try something different, especially when clients aren't cooperative. It's ok to mess up. When the game is over, you get to play again. Some will convert, some won't, so what, thank you, next. Those who believe and never stop hustling are winners.

Get on the Phone First

If you wait for your phone to ring, most agents will probably outsell you in your territory, even when you're investing thousands of dollars in business development and marketing budgets. To dominate, you can never rely on inbound calls only. Get more leads in the pipeline by getting on the phone first.

A Salesforce research conducted in April 2015 has shown that 92% of all customer interactions happen over the phone, with 85% of customers reporting dissatisfaction with their phone experience. So, fire up that dialer, or pick up the phone and dial. If you are one of those agents that need the motivation to dial, then you might need to reconsider your real estate sales career. If you think about it, your life depends on this step. Thus, to thrive in real estate sales, you need to get on the phone!

Set a Minimum Daily Dial

Be persistent and set a minimum daily dial goal to get clients on the phone. Research shows it takes an average of 8 cold call attempts to reach a prospect and 18 calls to convert. Dial today or use one of your favorite dialers to make a minimum of 30 calls per day.

Take Control and Be Direct

People will take more actions when you are direct with them! Being kind is excellent, but you being kind doesn't always get you conversions. You need to be straightforward and take control of the conversation. Top real estate agents are confident, and they direct the discussion to where they want it to be; a conversion—that is a scheduled interview on their calendar. There is always one person who's in control. Whoever is in control, wins. Get straight to the point. Set more appointments by being linear and direct. Activate more clients by taking control of every conversation.

Make Prospecting a Regimen

Make setting appointments a hobby. Setting appointments is the number one top priority for top agents. It is ingrained in their minds as a part of their daily routine. It is the one activation metric that makes all the difference in the growth of their real estate business.

Every day should be a prospecting day. Several sales surveys show that Thursday is the best day to prospect. Wednesday is the second-best day. However, don't let this stat stop you from prospecting on Mondays, Tuesdays, Fridays, and the weekends.

Many sales reps make the mistake of calling during lunch hours. Statistics show that the best time to call is early in the morning (between 7 and 8 AM), or later in the afternoon (between 4 and 5 PM).

Be Yourself

Avoid the sales voice and sales cadence. You need to lower the tempo and sound like a normal person. Be authentic, genuine, honest, real, and positive!

You are here to serve your clients. Smile and use a positive frame of mind when interacting with others. Research shows our brains work up to 31% better in a positive frame of mind.

No matter what, DO NOT be negative, defensive, complain, argue, push to sell, take things personally, or become emotionally affected.

Be Sold

Be confident by seeing yourself as the expert in your space. If you can't sell yourself, you can't sell your clients! Be confident in your abilities and be sold on what you say.

Be Bold

Be bold and have a contingent back-up, "Plan B." When clients try to get you off the phone or ask you to call back: *"You know Sally, I appreciate you asking me to call you next week, but would it be okay if I took 30 seconds to tell you why I was calling, then we can decide if it makes sense to hang up and call back next week?"*

Be Available and Responsive

A 2018 survey by *The National Association of REALTORS®* (NAR) shows that 30-40% of people use real estate agents that respond first and quickly. Please make it a habit always to answer the phone. It would be best if you never were too busy for a potential client. Also, answer all your clients' questions. There is no such thing as a "stupid question!" Every concern is paramount and an opportunity for you to build rapport.

Discuss Clear Next Steps

Always strive to close and lock in a clean next step. In most cases, it is a time slot on the schedule, a face-to-face appointment, or a required signature. Make sure you confirm that after each agreement.

For example, here's how closing the appointment step would sound: *"Bob, may I make a recommendation? How about I come in and meet with you tomorrow at 6 PM so we can dig more deeply into the benefits and the values I can bring to the table?"*

Here's how closing the confirmation step would sound like: *"Okay great, I will email you a calendar invite right now. Bob, since I have you on the phone, would you be kind to confirm that you've received the email?"*

When prospecting, pre-draft calendar invites where you would only need to enter the date, time, and client's email, and then hit send. Top agents have their entire prospecting and sales content pre-drafted, including calendar-invite templates ready before every phone call they make. They are prepared to interact, send the invitations, and ask potential clients to accept them while they have them on the phone.

Go After the Big Fish

Top real estate agents go big and chase more significant opportunities! You need to pivot and think big. Have the guts and the courage to aim high. Target people with astounding job titles and higher home values. Double your pipeline volume not just by the number of deals, but also by the size of each transaction.

Great opportunities take almost just as much time and work as small ones. Thus, segment your client list, and then dial those with significant potentials and higher home values first.

Practice Your Scripts

I am sure this is not the first time someone exposes you to this. Practice, practice, practice! Great actors are so great at practicing their manuscripts when they're acting. They look real; we forget it is a show. To become great at real estate, practice your scripts daily. Roleplay with your spouse, friends, and colleagues! Master what you say, when to say it, and how to say it. Practicing your scripts makes you flow naturally, build rapport quickly, and set appointments smoothly. The more you practice them, the more you will close.

Follow Up

The money is in the follow-up! Don't give up on the first call. Get ready to make about 18 phone calls to convert each client, not one, not two, or three, 18! Research shows that 80% of sales require a minimum of 5 follow-up calls after meeting in person. Studies also show that 44% of sales reps give up after one follow-up call. Don't let this be you!

Understand Prospects Needs

Engage your clients and get them talking. Only 13% of people believe a salesperson can understand their needs. Your job is to ask, listen, and understand your client's needs.

Quickly introduce yourself and get into the purpose of your call, and what matters most to your clients. Listen to learn about their needs, wants, and desires. Show that you understand their pain points.

Don't pitch! Instead, ask open-ended questions and address their problems. Get them to talk about their challenges. For example, *"Sally, I've been noticing many homes in your area are selling fast but not your house! Can you please tell me why you think that is? I'm all ears!"*

Ask for Referrals

Most agents immediately move on and get off the phone as soon as they come across, non-qualified clients. I highly recommend you spend your time only with qualified clients, but since you're already on the phone, you might as well ask for referrals.

When you find out prospects aren't qualified, shift your conversation to see if they have people in their network that might be interested in buying or selling in the next six months. 91% of people say they'd give referrals, but only 11% of salespeople ask for them.

Follow the "Yes" Ladder

A recent study published in the International Journal of Research in Marketing showed that "the frequency of people's compliance with a request can be substantially increased if the requester first gets them to agree with a series of statements unrelated to the request but selected to induce agreement."

Robert Cialdini, Author of *Influence: The Psychology of Persuasion*, is hailed as the "go-to book" on persuasion selling over three million copies and making Fortune's "75 Smartest Business Books." It's been a game-changer for the sales industry. One of Cialdini's six fundamental principles of persuasion is "commitment and consistency." The way to build this commitment and consistency is through the "Yes" ladder.

The "Yes" ladder is a persuasion method aimed at getting your prospect to say yes to a specific question or situation (setting up a meeting or making a sale). The process starts by getting them to say yes to a series of questions that begin trivial (where they are practically guaranteed to say yes) and become less trivial as you ask each question.

Follow the "Yes" ladder in your conversations to get clients to commit to small incremental decisions. Each subsequent "yes" they respond with makes them more likely to comply with the next, more significant question. If they object, never fight back or argue. Address objections early in the conversation by asking more "yes" questions. Most importantly, never put pressure on clients to take massive action logically.

Build Rapport

Be empathetic and build rapport with your customers to earn their trust and their business. Build rapport through emotional intelligence on both sides of the table: agents and prospects. Manage not only your emotions but also those on the other side of the call.

Use emotions versus intellect. People decide with passion and feelings; not with logic or intellectual means! You build rapport by making them feel better.

Persuade your clients to answer "yes" in their minds and to their questions. *"Do I like this real estate professional?" "Do I feel like I can trust this real estate agent?" "Did what this Realtor just say, solve my needs, and make me happy?" "Does she have my best interests at heart?"*

On the other hand, how do you build rapport with a frustrated client? Take control of the conversation and positively influence the situation. Otherwise, use "voice control," "match and mirror," "emotional intelligence," and other tactics to build rapport and show clients you care. For more on building rapport, see the "Credibility" section in chapter 3: DISCOVER YOUR CORE VALUES.

Keep Clients Engaged

Give your customers the illusion of control by implementing effective pauses. If they answer and pause, ask more relevant questions, and then break again. You want to get them talking. Allowing them the chance to speak will open opportunities to have a significant problem—you have to manage them talking too much.

Ask questions and do not pitch until you dig a minimum of three layers deep. Ask open-ended questions with a smile and a positive redirection. Always ask prospects to tell you more, using your pleasant and refreshing voice. Get to the heart of the conversation and uncover their motives; needs, wants, and desires. Dig deeper by asking a minimum of three consecutive questions about each point, topic, or need.

Here's a general example of simple 3-layered sales questions:

1. *What is it you're looking to fulfill?*
2. *How will you feel when you've accomplished it?*
3. *Would you work with me if I can help you achieve that?*

Let's say you are prospecting Sally, the potential seller! Here's an open-ended question that you can ask to test her motivation:

- *Sally, what have you thinking about selling your house?*

Moreover, here is how you dig three layers deep to understand her level of motivation even further:

1. *What will selling your house do for you?*
2. *What's critical about that to you?*
3. *Can you tell me more about that?*

Also, you can dig even more in-depth using the price of the house to test the willingness and seriousness of Sally's motivation:

1. *Sally, what price would you say would motivate you to make a move?*
2. *How much do you owe on the property?*
3. *May I ask what you will do with the proceeds from the sale?*

Dig even more in-depth using the client's emotional connection to the current area, and the new home. For example:

1. *Sally, what do you like best about your existing home?*
2. *What do you like least about your current home?*
3. *Will you want to stay in this neighborhood?*
4. *How would your ideal new home look?*

How about digging into the readiness of selling?

1. *Sally, when do you see yourself being ready to make a move?*
2. *Is moving within a specific time frame critical to you and your family?*
3. *Are there any circumstances that would enable you to make this move happen sooner?*
4. *Are there any obstacles that would prohibit you from making this move happen?*

Whether you are speaking to buyers, sellers, FSBOs, Expired Listings, or any other audience, asking the right questions to keep them engaged is crucial to setting your listing appointment. To keep your audience engaged, you must ask, listen, and comprehend. It is imperative to dig deep and ask a minimum of three qualifying questions in each area of interest. Touch on motivation, intent, readiness, likes, and dislikes. Not only you will control the entire conversation, but you will find yourself with a wealth of viable information that puts you at an advantage to close every potential opportunity.

Value First, Commission Last

Some agents negotiate commissions later in the prospecting conversations; others discuss them only during face-to-face meetings with clients. Nonetheless, it would be best if you only talked about commission once it is the only subject that is left to review before signing the agreement, after you have built rapport, and your client understands your values. Hence, this is why most successful agents delay the commission subject to the time they present their net sheet.

Utilize your MLS system, CloudCMA, First American's CostsFirst mobile app, or FidelityAgent ONE mobile app to quickly and easily create custom net sheets. Once your potential clients acknowledge your real estate values, you can then present your commission value, including estimates of costs and net proceeds. When you reveal perceived benefits before expenses, you will have a better chance of facing fewer objections and oppositions.

The FidelityAgent ONE mobile application is my favorite. It calculates associated fees based on each specific state and county. Plus, it is free.

> *Take the Next Step*
> *Execute Tactic 11.0: Dial 30 Calls per Day*

Practice Your Sales Scripts

Practicing your real estate sales scripts must be a daily activity. Roleplay with colleagues, family, friends, or your spouse! The more you practice, the less you'll mess up, the more natural you will sound, the more clients will trust you, and the more sales you'll close.

You can have thousands of leads in your CRM, but what matters is how many of them you can convert. So, it is not lead-generation that makes you money; it is lead conversion. Conversions occur from practicing, knowing your industry, and mastering what to say, when to say it, and how to say it. Have your scripts down, and you will undoubtedly convert most of those leads. Mess up, and you will inevitably lose them. Therefore, roleplay and religiously practice your craft.

Here are some high-level but straightforward steps to follow and impress your customers through your communications with them and the sales process:

1. Ask questions to understand your customers' pain points.
2. When asked, provide multiple options and solutions.
3. Explain the benefits of each solution you offer. Focus on what

matters to your customers.

4. Make sure your customers not only see but feel the joy of the result of each benefit.

5. You must always build rapport with customers to establish trust and credibility.

6. Use your phone not to sell, but only to follow up and to schedule face-to-face appointments.

7. Focus on building relationships! Always be communicating before closing (ABCBC).

8. Always strive to delight your customers with outstanding customer experience.

Here are three concepts to keep top of mind when practicing your scripts and communicating with your prospects:

*   Emotional Intelligence
*   Intellectual Intelligence
*   Pattern Interrupt

Emotional Versus Intellectual Intelligence

Use a combination of emotional and intellectual intelligence to communicate effectively and win your audience.

Emotional Intelligence (EQ) is the capability of individuals to recognize their own emotions and those of others, and then discern between different feelings and label them appropriately. EQ measures one's understanding of emotions and their ability to empathize and work with others. Use emotional information to guide thinking and behavior; manage and adjust emotions to adapt to environments or achieve goals. Emotional intelligence will help you engage more effectively and empathize deeply with your audience. Exert emotions to express empathy, intuition, respect, and understanding.

Moreover, *Intellectual Intelligence* (IQ) is a general mental capability that involves the ability of individuals to reason, plan, solve problems, think abstractly, comprehend complex ideas, learn quickly and learn from experience. Intellectual intelligence exhibits your competency in solving your customer's predicaments logically. Apply logic to ask qualifying questions and rate the customer's value.

Individuals with high EQ are better equipped to make use of their cognitive abilities. They possess the ability to inspire people to action and make others feel confident.

On the other hand, people with high IQ but low EQ sometimes

sabotage themselves because they cannot relate to other people. They cannot handle stress constructively, and they often find emotional connections challenging to maintain.

Furthermore, developing your EQ can help you access your innate intelligence and amplify your empathy, which can lead to career advancement and better relationships. Practice embracing your uniqueness and the uniqueness of others; Express your feelings and interpret those of your clients.

Nonetheless, understanding emotional and intellectual intelligence is necessary to lead your conversations with clients to appointments. They function in tandem and are both critical to personal development and sales success. Take advantage of the benefits both EQ and IQ bring to every sales conversation and connect with your clients on both levels.

Be Empathetic

Engage with clients through humor and fun facts when possible. Be considerate and empathize with both positive and negative life events such as a new baby, divorce, sickness, and more. This way, your clients always feel cared for and understood.

The real estate transaction can overwhelm your clients. Moving can stress them out too, especially if it is because of a repercussion or a significant life event. Instead of just stating you've been there, share a similar story on how it all worked out for you, past clients, or someone you know. Having been there and sharing previous experiences will work wonders in connecting with potential clients who are undergoing similar emotional stages.

Be Intuitive

When you speak to potential clients, they expect quick and informative responses. Be helpful and reference addresses, locations, and even public record data in your answers. Don't let clients' questions go unanswered. Otherwise, if you don't know the answer, be honest, and tell them you will get back to them once you have the correct answer.

Be Respectful

Don't be an assertive salesperson that pushes boundaries to make a sale. Irritating and displeasing potential clients will fire back negatively on your brand and reputation. Be respectful and understand that not every lead is ready to decide today. It would be best if you were flexible with people that are "just-looking" while still driving the conversation forward in a meaningful and respectful way.

Qualify Your Clients

While EQ is necessary to build rapport and connect with potential clients, IQ is crucial in qualifying them. Before getting to the big close, you must first determine if your clients are eligible.

Depending on your target audience, here is some necessary information you must get from your clients to begin the vetting process:

1. The type of target audience you're speaking to, such as buyers, sellers, FSBOs, or Expireds.
2. Their contact information, such as name, phone, email, and property address.
3. Their motivation level by finding out the timeframe they intend to buy or sell. Is it within a month, three, six, twelve, or more?
4. Find out if you're competing by asking them if they're speaking to or interviewing other agents.
5. If qualifying potential buyers, find out if they secured financing. Were they approved? Are they pre-qualified and working with a lender on potentially getting approved, or not yet?
6. Areas or neighborhoods clients are interested in living, price range, number of bedrooms, bathrooms, and other features or interests.

Once you receive the qualification information, use Intellectual Intelligence (IQ) to rate your client's value based on their quality and motivation levels. Assign precise qualification points so you can focus on the most qualified clients. For example:

1. Assign a 1-point score to potential sellers and 0.5 to potential buyers.
2. Assign 0.5 point to each lead with an email, and 0.5 with a mobile phone number.
3. Assign 1 point if the lead is hot (0-3 mos.), 0.5 point if warm (3-6 mos.), and zero if cold (6-12+ mos.).
4. Assign 0.5 point if the buyer is officially approved by a lender, 0.25 if conditionally pre-qualified, and zero otherwise.
5. Assign 1 point to each lead with a scheduled appointment time.

To illustrate, if you have a potential seller (1), that has an email address (0.5), a mobile phone number (0.5), that is selling in less than three months (1), and you have a scheduled appointment (1). This customer value has a 4-point grade.

If you have a potential buyer (0.5), that has an email address (0.5), a mobile phone number (0.5), that is buying in less than three months (1), also approved by a lender (0.5), and you have a scheduled appointment (1). This customer value has a 4-point grade.

With this scenario, you know that a client must have a customer value of 2 points or higher to qualify, or for you to consider it as a quality lead.

You can add other qualifying features such as geolocation to determine whether a potential client is within your territory, or use home price to pursue clients with higher home values. Such a grading system is flexible, and you or your broker should control it. It should act as an internal system within your real estate office to help optimize your conversion ratio. A grading system will save you time in your prospecting efforts as it allows you to focus on the most qualified leads generating the highest revenues in the quickest time.

To learn more about grading your leads, see "Lead Grading and Scoring" in chapter 10: NURTURE MEANINGFUL RELATIONSHIPS.

If you feel there's a need for a more straightforward grading system, prioritize your leads based on target closing date, separating them into three categories: selling or buying in the next three months, one year, or over one year.

- "**A**" = Less than 60 days
- "**B**" = Between 60 - 365 days
- "**C**" = Over 365 days

Regardless of whether you follow a lead scoring or grading system, you will need a lead qualifying questionnaire form to take essential notes while you're on the phone qualifying your potential customers. Create a lead qualification worksheet to interview and scrutinize your clients.

> *Take the Next Step*
>
> *Execute Tactic 11.1: Create a Lead Qualification Worksheet to Interview and Qualify Your Potential Clients*

Pattern Interrupt

Before you can convince or persuade anyone to do anything, there is one critical thing you must do first. Otherwise, you'll never have a chance of influencing or persuading anyone. What is that one thing?

Attention!

To get anyone to do anything, you must first get their attention. If you can't get someone's attention, then the likelihood of your ability to influence and persuade is virtually impossible. The most effective way of gaining someone's attention—especially when it is at such a premium because everyone else is trying to do the same thing—is through the use of pattern interrupt.

A pattern interrupt is your ability to stop a pattern and interrupt it by unexpected behavior. It is a technique that purposely goes against the norm to capture someone's attention. A pattern interrupt is a confusion technique to change an individual state or strategy.

Get Clients' Attention and Get Them Interested

A pattern interrupt occurs when a sudden insert of a word, movement, thought, sound, feeling, or visual element takes place. For instance, it can be in the form of an image, video, text, and conversational technique. The idea is to stop someone in their tracks by violating a typical pattern they are accustomed to so you can get their attention.

For example, a pattern interrupt happens when users are scrolling through social media feeds and stop because a post caught their attention. Something about that post interrupted the user's pattern of scrolling. Marketers often use the pattern interrupt technique to get their audience's attention.

Behavioral psychology and neuro-linguistic programming (NLP) use this technique to interrupt and change thought patterns, behaviors, and situations. It can be as simple as initiating a handshake or as definitive as seizing the moment to travel or fulfill your bucket list.

According to the National Science Foundation, humans have an average of up to 50,000 thoughts per day. Amazingly up to 95% are the same thoughts, repeated every day. That's too much repetition and very little space for new thinking. Einstein summed it up nicely.

> *"Insanity is doing the same thing over and over again and expecting different results."*
> —Albert Einstein

Turn your back on insanity and embrace pattern interrupts so you can get different results. As a real estate advisor, there are two pattern interrupt techniques that you can leverage to convert more leads to closed sales.

The Pyramid Pattern Interrupt (PPI)

Also known as "The Bottom-Up Pattern Interrupt." You ask questions to build rapport and eventually get to the top of the pyramid of the conversation.

Instead of customers treating you as a salesperson, you must break the pattern in the first eight seconds of conversation. Your first eight seconds on each phone call are the most crucial deciding factor regardless of whether they listen.

Catch your prospects off guard by using pattern interrupt. Get their attention, establish rapport, and show you are the person to whom they should listen. Customers need to feel your positive energy and enthusiasm. To be worth their time, you need to be fun and different.

Open in-person appointments and phone calls with pattern interrupts. Behave as you've already met the person before. Act as you know them to some degree. Ask interesting questions that somewhat throw the other person off. You want them to pause for a moment and process your request. You want them to be in the mental state of surprise, doing the opposite of what they expect.

Here are a few simple examples of pattern interrupts you can use to start your phone conversations while prospecting. (Read them out loud dramatically; they may make more sense.)

- *"Mr. Johnson?" "James from Dominate Real Estate gave me your number. He said you might need help with...selling your home..."*
- *"Hi Frank, it's Rose. How have you been?"*
- *"Hi Sally, Rose here from Dominate Real Estate. Have you heard of the new...listing in your community?"*
- *"Bob, did any of these ring true to you?"*
- *"Hey, if there is one thing you could do well in ...selling your home, what would that be?"*
- *"I got you! Well, hello, Barbara."*
- *"Hi Ellie, I was reading your post on Instagram about the topic, and it gave me an idea that might help you in..."*
- *"Hi, I was reading your profile on LinkedIn, and I noticed topic, and thought you might like what I have to say..."*
- *"Hi, the reason for my call is..."*
- *"Sorry, oh I'm sorry, can I just call you back in an hour?"*
- *"Oh, no! Sorry, Bob. Can you please hang on for just one second? Thanks!"*
- *"Wait...Please don't hang up!"*

- *"Wait, wait... I have something for you. Can I email you the-value?"*
- *"Hi Sally, I can't believe your house hasn't sold yet. I thought it was going to sell for more than it was listed! Just curious. What happened?"*
- *"Frank, with all honesty, may I suggest something helpful to you?"*
- *"Since we're both on the same page, what would you like to do next?"*
- Start your conversation with a relevant quote and ask, *"Don't you agree?"*
- *"I'm curious! Please help me understand why you say that Mr. Johnson..."*
- *"Bob, do the math..."*
- *"Sally, have you given up on selling your home?"*
- *"Hi, is there anything I can do to help you today?"*

When using pattern interrupt, don't mention negative remarks or use pessimistic statements. Here are a few examples that you shouldn't use:

- *"Is this a bad time?"*
- *"Did I catch you at a bad time?"*
- *"When would be a good time to call you back?"*
- *"I don't want to take too much of your time today..."*
- *"I am sorry to bug you..."*
- *"I know you're probably not thinking of selling or moving..."*
- *"I know you're probably tired of hearing from real estate agents..."*
- *"I am sure I am not the first real estate agent to contact you..."*
- *"I know you will not like this..."*
- *"Would you be offended if I asked for your contact information?"*

The Reversal Pattern Interrupt (RPI)

Also known as "The Top-Down Pattern Interrupt" technique.

Use this technique when intelligent customers try to pattern-interrupt your method of pattern-interrupting them. For example, when you're speaking to Sally the Seller on the phone, and she asks you to hold on while you're in the middle of your conversation, Sally just pattern interrupted you.

When Sally comes back on the line, do you start over? No, you do not! You do the reversal pattern interrupt technique. You start from the end. If you were trying to lead the conversation to an appointment, you no longer try to build that pyramid, you begin from the top. You ask, *"Let me make this quick: are you available to meet later today or tomorrow?"* Sally was not expecting such a question; neither was she prepared to answer.

When you pause and proceed with another reversal interrupt, *"I guess later today might be difficult; all right Sally, would you prefer 2 PM tomorrow?"* You took her by surprise. She might agree to meet.

If Sally the Seller says something like, *"I want to think about it,"* pattern Interrupt that with a smile and say: *"Sorry, Sally, you're not allowed to think about it."* Pause, now answer:

"Usually, when people say that to me, it just means you want to say no. I appreciate you being sensitive to my feelings, and thank you very much for that, but you know what? We are both adults. If there is a reason that is stopping you from wanting to do the deal, tell me what it is, perhaps I can help. Otherwise, say No, and it is over. I promise you won't offend me. Are you certain about not leveraging the opportunity of us working together?"

Again, Sally was not expecting such an answer. If she's in the business to sell, she will meet. Otherwise, ask if she knows anybody that's thinking of buying and selling. Thank her for her time and get off the phone.

Buyers Scripts

When you receive phone calls about a potential listing, have your scripts down to qualify your buyers immediately. There are usually two types of potential buyers that call. The curious buyer would possibly be driving around, kicking the tires, sees the "For Sale" sign, and dials the number. The caller may be a neighbor wanting to find the price of the house listed next door.

The next type is the serious buyer who is actively hunting for their next home. They are looking to make a move and see what's available. It would help if you qualified your potential buyers immediately to find out the curious from the serious. For example:

- *Bob the Buyer: Hi, I am calling regarding your property on 123 Main St. How much is it?*
- *Ross the Realtor: Are you just inquiring about the price or looking to buy within the next 30 to 90 days?*

If they are just curious, tell them the price and get off the phone. Otherwise, continue qualifying serious buyers.

1. Once you've asked if they are in the business to buy and they seem to be serious, ask about the kind of home they're seeking.
2. Ask if they're paying cash or getting a loan.
3. If they're getting a loan, find out if they've talked to a lender

about getting pre-approved.

4. If not, advise them to talk to a lender to find out how much they can afford. Explain why it is essential to discover how much they can afford, also, how an official pre-approval letter from their lender can give them an advantage in their offers. Provide a list of contacts for your preferred lenders via email or SMS.

5. Ask if they're already working with an agent. If so, with whom are they speaking, and did they sign an exclusive agreement?

6. Close buyers offline! Invite them to come into your office. Advise them that going through the home buying questionnaire in person helps you better determine the homes that best fit their needs. If not, try going to their home. Otherwise, you may agree to meet at the property they are interested in buying.

Many agents begin their relationships with their potential buyers by verbally asking the qualification questions, while others email the questionnaire form. Whether you vet your buyers on the phone or face to face, get them to sign an exclusive agreement with you. Neither be their virtual assistant nor their driver.

Here are sample questions to ask when qualifying potential buyers:

1. *Where do you live?*
2. *How long have you lived there?*
3. *Do you own or rent?*
    a. If they own: *Will it be necessary to sell your home first?*
    b. If they rent: *When is your lease due?*
4. *Are you employed? Where do you work? What about your spouse?*
5. *How many are in your family? Do you have kids?*
6. *Would you describe your present home for me?*
7. *How about the new home, can you tell me what you're looking for?*
8. *Have you seen anything you liked?*
    a. If they say, "Yes:" *What prevented you from buying?*
9. *How long have you been looking for a home?*
10. *How soon are you looking to buy? Can you please tell me why?*
11. *If we found the right home today, will you sign with me?*
12. *Will you be financing or paying cash?*

> ### Take the Next Step
> *Execute Tactic 11.2: Qualify Your Potential Buyers*

Sellers Scripts

There are many factors home sellers consider when they are contemplating making a move. Price is one of the most important. Second is timing, because when they want to sell is crucial. You can determine the date of the sale by when they want or need to move.

Once sellers have decided on a move date, you, the real estate agent, can then analyze the approximate time their home may take to sell. Look at the "average time on the market" stat for comparable homes in their price range. You can gather such information and regularly present it to the sellers during the listing presentation.

When your sellers decide on selling, they must view their home through the eyes of potential buyers. Come in and educate them on some points they must consider to increase their chance to sell their home quickly. These points help in making the home "show shape" or "show ready." Perhaps the house needs some improvements, minor repairs or certain upgrades, such as new kitchen counter-tops or a new carpet. Maybe it needs some new light bulbs, and simple paint touch-ups.

If there is anything the house needs, be honest and direct. Sometimes a few simple changes to the house go a long way. With the proper time, pricing, and presentation, you can help the sellers eliminate all the hassle. Openly disclose any necessary enhancements or adjustments that will help sell their home faster, maximizing your time and their profits, as well as yours.

Moreover, you need to find your sellers' intentions and measure their level of motivation. It would be best if you also asked a series of questions to qualify them before setting face-to-face appointments. Begin by asking general property information. You want to get them talking, open up, and be comfortable on the phone. Then, steer the conversation based on their needs. Here are sample questions to ask:

1. *Hi Sally, can I please confirm the address of the house you'd like to sell? OR*
2. *Hi Sally, you were just on my website. I understand you want to know the price of your home. Is that correct?*
3. *Have you made any improvements or updates to your home that would impact its value?*
4. *Great, what did you have done?*
5. *How much did you put into that?*
6. *Sally, can you please tell me more about your house?*
7. *Fantastic, and thanks for the information. To provide you with the*

most accurate home valuation, I would only need 15 minutes of your time. For us to both agree to put a correct price on your home, I would need to see it, right? OR

8. *Excellent, and thanks for that. I would need to look at the house and the improvements you've made to incorporate them into the home's value. It only takes 15 minutes. You want an accurate price for your home, right? OR*

9. *I received your online home evaluation request, which we offer free and with no obligation to list. To complete it, I need to take a brief tour of your home. When would be a good time for you? Again, this is a free service with no obligation to list.*

10. *How about we meet tomorrow or the next day? Whatever works better for you. Does this sound good?*

Note: If you are making an outbound call and they hesitate by asking how you got their information, you can inform them that their data is available through public records.

1. *Sally, don't worry, I'm not looking to sell you anything! I want to make sure I have the right information and that I am speaking to the right person. Your information is a public record I have in my system, but I am required to ask questions to protect your privacy.*

If they also refuse to meet or object by saying, "I am just curious, I am just looking, or I am not interested," continue the script.

1. *Perhaps a free listing alert might be just what you need. Do you know what that is?*

2. *Every time a neighbor lists their house on the market, I will notify you instantly via email about the listed price and their property details. The notice will give you an idea of what your home is worth. It is free, and you can customize it from your end. Does this sound like a report you would want to receive?*

If they don't cooperate, leave your information and get off the phone. Otherwise, proceed with follow-up questions to learn more about them, their property, and their intentions:

1. *Since I have you on the phone, do you plan on making a move in the next 3 to 6 months?*

2. *How many beds, baths, appealing features?*

3. *Do you still have a mortgage? How much do you owe?*

4. *Are your taxes current? Are there any liens or judgments?*

5. *Is the house listed with another real estate agent?*

Now, understand the seller's motivation. You want to uncover their real motivation politely. If you don't get the information you need, then follow up with these questions until you get all the answers:

1. *Fantastic! Can you tell me about your situation? I'm just wondering, why are you selling?*
2. *Just to clarify, would you say you really 'need' to sell, or maybe just 'want' to sell?*
3. *How quickly would you like to sell this house? Are there any pressing time constraints?*
4. *What is the lowest you would go with a cash offer?*
5. *Do you have any obstacles not to sell?*
6. *What are your backup plans if you don't sell? Have you thought of any?*

Learn more about the property and its condition. You want to get clients talking about their property and what they think of it. If you don't get the answers you're looking for, use these follow-up questions:

1. *What's the condition of your house?*
2. *Does it need any repairs or maintenance?*
3. *Does it need other work? How's the condition and age of major systems, such as the roof, HVAC/Heat/AC, water heater, plumbing, electrical, flooring, paint, counters, appliances, windows, structural condition, or cracks?*

Understand upgrade costs and property value from the seller's perspective:

1. *What do you think your house would need to bring it up to tip-top shape to get you the most money?*
2. *If you had to guess, how much do you think that will cost?*
3. *What do you think the property is worth?*
4. *How much do you want for it?*

Now, go for the close:

1. *Great! Thank you so much! I appreciate you answering my questions. Let me review this information and get back to you within the next 24 hours. May I reach you back in case I have additional questions?*
2. *OK, great. What's the best way to reach you?*
3. *Is there anything else you can think of at all, that I should know, or take into consideration before doing my research and getting back to you?*

4. *Wonderful! When would you like to meet and go over the details of your options, today or tomorrow?*
5. *Will 10 AM, 2 PM or 6 PM work for you?*

Take the Next Step
Execute Tactic 11.3: Qualify Your Potential Sellers

FSBOs Scripts

Just like with every prospecting call, you want to do a little research before calling a potential lead. Then, leverage gathered information in starting a pleasant and personable conversation. Research is especially useful when calling a potential client for the first time.

Before you call a For Sale by Owner (FSBO), find out where sellers are promoting the property. FSBOs typically list their houses on some of these websites: Zillow, Home Finder, For Sale by Owner, FSBO, FSBO Local, Home Bay, Hot Pads, Let Go, Fizber, Craigslist, Asset Column, Find My Roof, Oodle Marketplace, Facebook Marketplace, LoopNet, or local advertising sources. Note, you can also use these websites to find your free FSBO leads.

When you make the call, immediately make sure you're speaking to the owner, if not, ask to talk to the owner of the house for sale. Quickly introduce yourself and inform the owner where you found the listing. Once you get their attention, immediately ask away.

The introduction calls between Rose the Realtor and Frank the FSBO would go:

1. *Hi, I am looking for the owner of the house for sale. My name is Rose with Dominate Real Estate. I found your home listed for sale on **FSBO.com**. I wanted to find out a little more information for qualified homebuyers with whom I am working. Do you have a few minutes to answer some questions about the property?*

First, you need to know if the house is still for sale:

2. *I work with many buyers in the area, and I am committed to finding them the right home. Is your place still for sale, or are you thinking of selling?*

Then gradually build rapport by asking simple questions about the house. Again, you want to get them talking:

3. *How many bedrooms, beds, baths, kitchen square footage, and stories?*
4. *Does it have a yard or a pool?*

5. *Did you make any repairs or upgrades?*
6. *What do you like about the neighborhood?*
7. *What is the current sale price?*
8. *How did you determine the sales price?*
9. *Have you had a recent appraisal?*

Then, be curious to measure demand and the number of current offers:

10. *What's so special about the house?*
11. *I am just curious; about how many calls have you received on this property in the last seven days?*
12. *How many of those calls were from real estate professionals?*
13. *Are you saying there was only about "X" buyers who called?*
14. *How many of those people came to see your house?*
15. *How many offers have you received from them?*

Ease the conversation into measuring intent and motivation levels:

16. *It sounds like you have a great home; may I ask why you're selling?*
17. *While I have you on the line: May I ask where you are moving?*
18. *How soon do you need to be there?*
19. *How did you decide on the area?*
20. *Are you buying a new house? How much?*
21. *Do you have to sell this first? In what time frame?*
22. *How long have you been trying to sell your house by yourself?*
23. *Are you willing to drop your price when working with qualified buyers ready to buy now?*

Now, elevate your values as a real estate advisor:

24. *From what kind of buyers did you get the offers? Can I explain?*

25. There are only four (4) types of buyers out there:

   a. *The Serious, and in a Hurry: Mostly Relocators. Those are the earnest buyers who are moving or buying a replacement home. Those are usually in a rush to purchase a new home due to either their existing home is selling soon or have just sold. These buyers almost always have an agent helping them. These are the best type of buyers. They have the money, and often, not the time. They don't look at FSBO ads because they want an agent to find them the best home fast. Their agents tend to work with other agents for many reasons. 97% of national listings are with real estate agents; all other listings account for only 3% of the market. Where would you think an agent would search for homes, in the 97% inventory or the 3% where FSBOs and others list? Furthermore, working with other industry professionals leads to a faster closing. That is what is important to these buyers. Does this make sense? You would want offers from these serious and in a hurry type of buyers, right?*

b. *The Serious, but Not in a Hurry: Mostly First Time Home Buyers. Those are committed buyers but aren't in a hurry to buy. They look at on average of about 45 homes before deciding. According to Genworth 2018 Research, they account for 55% of the market. They usually need financing and handholding through the process. Naturally, they hire real estate agents to negotiate the best deals for them because they want to buy homes ten cents on the dollar. They also need agents to represent them in the home buying process. Do you think their agents would want to work with their fellow agents or call a FSBO? You would wish to receive offers from these committed buyers, right?*

c. *The Lookie-Loos: Mostly, browsers. Those are curious people who look at homes for sale without intending to buy. They usually browse alone because most licensed real estate professionals don't represent them. These people are typically not qualified to buy. You wouldn't want an offer from these buyers, right?*

d. *The Investors: Mostly, they're bargain hunters. They prey on foreclosures, expired listings and the For Sale by Owner. They are usually in a hurry to buy and sell at a discount. You don't want to sell your house at a discount, right?*

26. *Where would this leave you, Frank?* (Frank now realizes that he must work with buyers represented by real estate agents, not the Lookie-loos or the Investors).

27. *Now, out of those "Y" offers you've received, how many were from serious buyers?* (Frank soon realizes that he might need to hire a real estate agent as these committed profound buyers aren't out looking at FSBOs).

28. *What type of marketing are you using to sell your house? How are you marketing it?* (Frank now realizes that he needs a solid marketing plan to sell his house).

29. *Would you like to know about our Marketing Guarantee and all the marketing techniques I use to get your house sold?* (An "aha" moment! Frank now realizes that you have the values he needs. The benefits of hiring a real estate agent he wasn't aware of).

Qualify even more and shift into closing questions to see if there is a potential in getting an appointment:

30. *Are you cooperating with agents on the sale?*

   a. If not: *Well Frank, you're stuck working with The Lookie-loos and The Investors. You said you don't wish to work with them, right?*

31. *If I bring you qualified buyers, and you get the money you want, will you pay me a commission?*

   a. If you're going to be specific about your commission: *Will you be willing to pay me a 3% co-op? Alternatively, the 3% broker commission?*

32. *Realistically, how long will you continue to try and sell on your own before you decide to explore other options?*

33. *In that time frame, if you decide to hire an agent, do you have someone in mind you trust like a close friend or a family member with whom you feel obligated to list your house?*

34. *What's the likelihood of you interviewing and working with other aggressive agents like me?*

35. *If I can get you more money, will you consider using me as your real estate agent of choice?*

36. *Wonderful! If I can sell it for you, you have the option to take the offer and pay my fee. However, if you sell it on your own, you don't owe me anything at all. Does that sound like a fair deal you would consider?*

37. *Great! I'd love to come by and share more ideas on how to sell your house fast. Would tomorrow at 2 PM, or 6 PM work for you?*

38. *No, no worries! What works better for you usually, mornings, afternoons or evenings?*

39. *Excellent! You're on my schedule for tomorrow at 10 AM. I am sending you an email invite right now. Can you please confirm receipt?*

If answers don't align with your requirements, delete the lead, move on, and don't waste your time.

For example, if a Frank isn't willing to pay a commission, don't even bother proceeding with the other questions. *"Thank you for your time, Frank! Here's my information in case you change your mind..."* Hang up and move on!

Otherwise, if he's willing to pay a commission, find out the waiting period. Preferably, and ultimately, you want him to say he is ready to list with you right now. Commitments will not happen often, but if you call enough FSBOs, they will happen. So, create a set time frame preference as your intention on every phone call you make to a For Sale by Owner

(FSBO). For example, a maximum of 30 days or 60 days, you decide! Then, follow up to set the appointment when they're ready.

If Frank's timeframe works for you, find out if he's willing to hire an agent. If not, again, leave your information and move on. Otherwise, continue your script to accomplish the two most important objectives:

1. Set a home preview appointment, and
2. Set a legitimate listing appointment.

Moreover, each FSBO must meet the above qualifying parameters before you want to preview the house. Nonetheless, a preview appointment is when you go out to look at their place of residence and meet with the sellers for a few minutes. Your purpose is to build a healthy rapport in person. You will also need to genuinely dig into their motivation so you can move to the next step and set the listing appointment. Focus on qualifying your FSBO again while on a tour of the house. You'll likely receive more truthful answers when speaking face-to-face.

In close, thank Frank for having you over and ask if he would consider listing the home with you immediately.

At the preview appointment:

"Frank, thank you so much for your time today. I appreciate you giving me a tour around the house. Now that I have seen your home, I am confident we can sell it because of the point out the positive values in the house. Would you like me to make honest suggestions I believe will get you even more money?"

Point out any negatives, fixes, or upgrades needed to sell the house.

"Now, you see how transparent I am and how much value I can bring to the table. If I can sell your house for the desired price, in the desired time frame, and without you having to deal with all the hassle, will you consider listing with me?"

This closing question is critical to ask before you leave the house. It will typically elicit one of two responses. If Frank says, "Yes," set the appointment to come back for a listing presentation. If he says, "No" or raises a commission objection, also set the interview date.

"Frank, please allow me to go back to my office and do my research. I have to crunch the numbers and come up with a few price options for you. I would only need another 15 minutes of your time. Will tomorrow at 10 AM again work for you?"

Make sure to download the FidelityAgentONE mobile application. Use it in your listing interviews to show FSBOs that hiring you as an agent yields more money to sellers.

What to Do After the FSBO Home Preview?

Once you go on a tour with a For Sale by Owner (FSBO), a massive lead follow-up is now essential. If Frank will list within one week, call him every single business day and ask for the listing. If he's more than a week away, call him at least three times per week. The follow-up calls are easy because Frank now knows you. Therefore, you did separate yourself from your competition.

Besides calling Frank, consider following up with him on other channels such as social media and direct mail. Send him direct messages online and postcards in the mail. Be aggressive and send at least 2-3 mailers per week. Do not let him forget about you! The more you follow up, the more listings you will take.

> *Take the Next Step*
> *Execute Tactic 11.4: Prospect and Qualify For Sale by Owners*

Expired Listings Scripts

Expired listings are different as they are tougher to close. You need to pursue more of an aggressive approach to get them in dialogue on the phone. They don't want to speak to an agent. They are typically upset and frustrated. The previous agent didn't do the job needed to sell their home. Whatever it is, you need to warm them up progressively.

At first contact, you need to dig deep and get them to open up by asking open-ended questions.

Empathizing and building rapport are critical when speaking to sellers with expired listings. Lead them to "self-discovery" of their pains through your conversation with them. No matter how the conversation goes, offer to solve their needs through your guarantees and your marketing plan, not you.

Ask questions around what went wrong to uncover viable solutions. Ask for their opinions to hear their perspective on your "soon to be proposed" solutions. Feel them out, hear them out, listen to what they are telling you, not just what they are saying. Feel their frustration, see through their eyes, and put yourself in their position before you pitch any offers or solutions.

Once you feel you've built enough rapport and listened to their frustrations, steadily qualify them, quickly move on to illustrate your values, and then arrange the meeting.

Here's a quick FSBO script by Rose the Realtor speaking to Ellie the Expired:

1. *Ellie, I saw your home came off the market, and I wanted to find out if it is still available or if it sold.*

2. *OK, so you're telling me that the last agent did not do much to market and sell the property.*

3. *Your home sat on the market for six months with no viable offers.*

4. *Wait, you only had two showings, and you needed to get the property sold last month.*

5. *Ellie, can I be direct with you?*

6. *Listen, Ellie, that is the reason we have four guarantees in place. They are in effect to protect you. We specialize in selling homes that didn't sell the first time. We've put measures in place to avoid devastating situations like this from happening at all costs. When a home doesn't sell, it doesn't just adversely affect you as a client, but it also hurts us and our reputation as a real estate brand. It's terrible for everyone involved. Don't you agree?*

7. *Would you like to learn more about our guarantees and how I can get your property sold quickly and for the most money?*

Move on to the close; otherwise see Communication Guarantee, No-Risk Guarantee, Performance Guarantee, and the Marketing Guarantee in chapter 3: DISCOVER YOUR CORE VALUES. Drop teaser lines to elaborate briefly on the guarantees if asked, and then close.

8. *Ellie, we are professionals, and we take customer service seriously. Every one of our clients gets update calls on Wednesdays and Fridays. Even when we have no new updates, you'll still hear from us. Besides, we guarantee a 60-day close. If you're not happy with us in the first 30 days you list with us, you can cancel with no obligations. We have a thorough and concrete 75-Point Marketing Plan that gets your house sold fast! If there is anybody that will get your property sold, it will be me.*

9. *Ellie, I would love the opportunity to sit down with you and discuss our guarantees, real estate services, and your needs in detail. Does later this afternoon or tomorrow morning work for you?*

Don't just adhere to scripts outlined in the book. Understand your clients' needs, wants, and desires before closing. Progressively ask qualifying questions to achieve a listing appointment. Practice your craft and stay focused on the path to close. Observe how these series of inquiries progress to closing:

1. *Ellie, do you still feel like you need to sell your house?*
2. *How fast would you like to sell?*
3. *Any particular reason why?*
4. *Ellie, may I ask where you are you moving?*
5. *How soon do you need to be there?*
6. *I'm curious; why did you first list your home for sale?*
7. *Ellie, what made you go with the last real estate agent?*
8. *Did the last agent do anything you liked or impressed you?*
9. *What did you not like about the last real estate agent?*
10. *Ellie, why do you think your home didn't sell?*
11. *If you were the last real estate agent, what would you have done differently?*
12. *If I can save you the hassle and do all that for you, will you work with me?*
13. *Great! Are you usually available mornings, afternoons or evenings?*

When working with sellers, FSBOs and Expired Listings, the goal is to get them to say "yes" to these types of questions:

1. **Propose Your Values:** *If I was to save you the hassle and bring you all these values, will you work with me?*
2. **Present Your Services:** *If I could bring you an offer that would lead to an acceptable net price, would you accept such an offer through my real estate firm?*
3. **Proceed to Close:** *Can I come in for a few minutes to see your home so I can tell potential buyers about it?*

If you receive any objections to these three major steps, go back and ask more qualifying questions so you can understand why they're hesitant to work with you.

Scripts keep you focused on your activation metric—the conversion. Please do not follow them word for word. Use them to help you get better at engaging with your potential clients, not to sound like a robot. Practice your craft to keep the conversations short and relevant. In the end, stay naturally conversational and do not digress. Be yourself and use common sense.

> *Take the Next Step*
>
> *Execute Tactic 11.5: Prospect and Qualify Expired Listings*

Now that you have learned about emotional intelligence (EQ), intellectual intelligence (IQ), the different types of pattern interrupt techniques, as well as have gone through samples of detailed sales scripts for buyers, sellers, FSBOs, and expired listings; use what you've learned so far to get in the habit of practicing your sales scripts daily!

> *Take the Next Step*
>
> *Execute Tactic 11.6: Practice Your Real Estate Sales Scripts*

Disclaimer

Make sure you and your real estate broker adhere to Federal Laws and the Do Not Call (DNC) Registry. Purchase the DNC list or hire a professional compliance service to cross-reference consumer information before contacting them. (See "DNC" List section in chapter 7: BUILD MASSIVE MOMENTUM).

Handle Objections

What stops people from wanting to work with you—also known as Objecting—is one factor, "You." You have not shown them your values and their benefits; hence, they resist. People will pay top dollar to premium products and services. That is because premium brands unveil their unique values. For example, when you walk into an Apple Store or Nordstrom, do you object to the price of their products? Some stand outside Apple stores the night before a new ingenious gadget is out.

It doesn't matter if you work with top real estate brands, objections will occur if you haven't proven your values as an agent. The real estate business is personal. People want to trust that you have their best interest at heart, and that you will get them the best deal. When they object, it means you have neither shown them the values they're looking for nor that you are the best agent in town.

If you come across potential customers who raise objections often, go back and apply all concepts explained in chapters 3, 4, and 5. Nevertheless, here are some common objections most real estate professionals face and how to deal with them.

How did you get my number?

You know Bob, our technology team pulled this list together for me. Somehow your number is associated with this house address through public records.

- If they are upset: *Bob, I understand. I will remove your number right now, and you won't receive another call from me.*
- If they are just curious: Continue to ask questions.

I already have an agent

Have you signed a contract with your agent yet?

1. If no: *Great. I understand you want to work with someone with whom you are already familiar. Just out of curiosity, with whom are you working?* Get the name of the agent, and continue your script.
2. If yes: *Are you thinking of switching agents?*
   a. If they are, continue qualifying your prospect. Otherwise, get off the phone.

I have a friend who will give me a discounted commission. Will you do the same?

Sally, real estate agents who cut their commissions rarely have all the skills and the resources top professionals like me have. Assuming your friend does, are you telling me he/she will go above and beyond to do all the extra work I'll be doing for less commission?

Besides, which agent do you think will negotiate more aggressively on your behalf, the agent that is not willing to negotiate with you, and yet accepts a lower commission, or the aggressive one?

Also, which agent do you think will work harder and get you the most money, the one that chooses the easy way out, or the agent that stands by his/her values?

Can you cut your commission?

Sally, can I explain to you how my real estate commission works?

1. Pull six $1 bills out of your briefcase, wallet, or purse.
2. *Sally, I've got six $1 bills here, what most people don't know is three of them automatically go to the agent on the other side.* Set three of them aside.
3. *One goes straight to my broker.* Set that one aside too.
4. *One goes to cover all my business expenses, including my license fees and taxes.* Set that dollar aside.

5. Hold up the last dollar. *Half of this goes to cover my continuing education.* Now, rip the dollar bill in 1/2.

6. *Sally, do you want to take my half of a dollar bill?*

I'm not buying for another three to six months

That is great, Bob! Can I tell you why? Since you're not in a hurry to buy, it would be an excellent time to do some research together. I will have plenty of time to understand which home you see yourself living in. We will also be at an advantage in negotiating the best deal. You want the best home at the best price, right?

I don't want to commit to one agent at the moment.

I completely understand. It is normal, and you may interview more than one agent. It is your prerogative to find the agent who stands out and looks for your best interest. Based on the values I am bringing to the table, and everything else I have shown you, do you feel like I can (if speaking to a buyer—*find, negotiate and get you into your new home*) or (if speaking to a seller—*list, negotiate and sell your house*)? *Great! Then, let's execute this exclusive agreement.*

What if things don't work out between us after I sign the contract?

I understand your concerns, and I appreciate your honesty. In my experience, this only happens because of miscommunication between both parties. That is why we have a Communication Guarantee policy that promises an open, honest, and transparent communication from our end.

We also have a No-Risk Cancellation Guarantee policy where you may cancel with no cost or obligation to you within 30 days if you are not happy with my services for whatever reason. These two guarantees will protect you should things not work out between us. Let's get this contract signed.

My listing agent is giving me a point back if I buy a home through her. Will you do the same for me? If not, why should I use you?

Are you saying you will risk selling the house for one point? Your listing agent is busy trying to get your existing home sold. She can't possibly focus 100% on selling your home, and 100% on finding you a new home, right? There will be a sacrifice in both processes. The risk will fall on you. On the other hand, I am 100% focused on finding you a new home. Also, an additional reason you should work with me is that the other seller pays my commission, not you! Does this make sense?

I only want to work with a large real estate firm

May I ask why? Are you saying you only want to work with a significant real estate firm with many agents, so you have a broader exposure? I have a solid marketing plan that reaches fellow real estate agents in all companies, your county, and the surrounding three counties. Top producing agents in big firms leverage their network. I am one of them, and I am well-connected too. For example, I attend broker previews weekly, and I send out digital flyers of listings to thousands of contacts in my database. My marketing efforts are far more powerful than having to work with a large firm. Do you see the value in that?

Please keep in mind that when a prospect becomes agitated and continues to question you, your ONLY response should be humble, apologetic, and assuring that you won't contact them again. Never attempt to engage in debate with an upset person!

> **Take the Next Step**
> *Execute Tactic 11.7: Practice the Handle Objections Scripts*

Lead Follow-Up

Research shows that the top two consumer complaints for the last ten years have been communication and follow through. On every first meeting with every new client, determine the best ways to converse with them. How would they like you to communicate with them when you follow up? Do they prefer you contact them via phone, text, email, or social media? How often would they want you to follow through? In this section, we will explore The Three F's for communicating with your database.

Follow Up

Follow Up with your leads after prospecting, also, with the opportunities that might turn into deals. Follow up with deals that might turn into clients, with new clients, past clients, partners, friends, and family. If you haven't noticed, your entire business is dependent on following up. Hence, why Tom Ferry keeps stressing, *"The money is in the follow-up,"* and Ryan Serhant takes *follow-up* two steps further, discussing the importance of *following through* with clients and *following back* with past clients and cold leads.

Real estate professionals and firms invest large budgets in creating demand, generating leads, prospecting, and converting sales. Following up in each step of the process is the "insurance policy" to continuous momentum and thriving business. Out of sight, out of mind! The more you check-in and check-back with your relationships, the more likely to close more sales.

How do you follow up? On every channel possible! In chapter 7: BUILD MASSIVE MOMENTUM, you prepared your farm for growth. You collected and verified contact information such as property addresses, email addresses, and mobile phone numbers. Also, you obtained social media profiles such as Facebook, Instagram, and LinkedIn. If you are only making one or two contact attempts via phone before trashing the leads, you are losing out on many opportunities! In your follow-up process, you need to incorporate push notifications, text messages, phone calls, voicemail drops, email drip campaigns, and one-on-one emails, social media direct messages, chat messages, and direct mail. Follow up using every communication channel available to you and your clients.

If you can communicate and follow up with your sphere on all these channels, worry not. You will not annoy them; you will impress them with your omnipresence. They will probably hire you because you will come off like you're on top of your real estate game. Just don't push the boundaries. Manage the follow-up sequence depending on which stage clients are. A good rule of thumb is to follow up three times a week.

Maximize your chance to connect with people. Do not let relationships cool down. Build a follow-up system within your CRM that automates your reach out and alerts you when you're required to follow up. You need a system that keeps track of all efforts. Break down your leads to hot (ready), warm (thinking of buying/selling), and cold (not actively looking). Set up automated follow-up plans for all leads, especially the warm and cold ones — also, design dynamic follow-up email templates and pre-drafted text campaigns. For example, create a basic newsletter email template that shows the recently sold, pending, and active listings in your clients' neighborhoods. Schedule to send email and text drip campaigns automatically to potential clients you've been prospecting.

Follow Through

Follow Through with other campaigns that provide personalized campaigns to warmer leads. Record voice notes or videos and send them via email, text, or social media. This way, you can build rapport, trust, and get that listing appointment set up or the contract signed.

The real estate business is personal, and customer service is paramount. Treat every potential customer as a new client. Do what you say you will do. Be dependable, reliable, and accountable. Follow through and go beyond expectations.

Follow Back

Follow Back according to the different emotional states your clients go through. For example, the initial sensitive stage, Bob goes through when buying a new home is "Excitement." At this stage, you can feel Bob's joy and enthusiasm in every communication you have with him. Bob might think he can afford double what he realistically can. In your follow-up process, your job is to qualify and move him to the home search step. At this stage, you might need to continuously and positively influence Bob because of his frustration finding the right home.

When it is time to sign the offer, fear might be Bob's best friend. Here is where trust and benefits in helping Bob determine whether he is making the right decision is vital.

Also, Bob will be in the stage of disappointment when his offer doesn't get accepted. Then, acceptance when he is finally going through the buying process. Happiness when closed! Finally, Bob will feel a sense of relief after a few weeks of settling in his new home.

Knowing what stage your clients are emotionally experiencing is invaluable for arming yourself, and your business with the best follow-back plan.

Consider following back with past clients and even lost opportunities. Be present and be there for those who are not in the market to buy or sell. Eventually, they will qualify to be potential clients. Don't just delete the leads who may not seem interested in your real estate services in the next 12 months. Have a follow-back strategy, so you don't miss out on any opportunities. Even if you planned to touch them once a month or once every quarter; circumstances change, and they change often. Open the floodgates to your real estate business and follow back with everyone. You never know who will be your ideal next customer. The best ones are sometimes the least expected.

> ### Take the Next Step
> *Execute Tactic 11.8: Follow up with Your Database*

CHAPTER TWELVE

SEAL THE DEAL

"Sales solves all problems."
—Mark Cuban

In the previous chapters, you worked hard setting up your real estate brand; you developed your websites, brand content, and lead generation tools. You were challenged! You were outside your comfort zone, but thankfully, all lead nurturing, and conversion campaigns are in place. In this chapter, although you'll feel like you belong; don't take concepts lightly as this is where the rubber hits the road.

Lead Monetization

A real estate agent monetizes when and only when a deal closes. There's usually a battle to get the contract signed, but there's always another to seal the deal and get paid. Don't leave securing your commission check to the operations department or anyone else. Concluding each real estate transaction is solely the agent's responsibility. Closing each sale is predicated on how the real estate professional handles each transaction from the first phone call to appointments, consultations, presentations, and through each step of the process. You control the path to gain or pain. Hustle, hone your skills, hold yourself accountable, and strive to earn what you deserve!

Working with Sellers

You've optimized your entire real estate business to create awareness and generate seller leads. You nurtured those contacts and persistently followed up with them. You eventually got them to agree to a listing presentation. Now it is time to showcase your expertise and reveal your track record. Share your company's sales history, and use your value proposition to win over those sellers and earn their business.

The Listing Presentation

The more you prepare for each listing appointment, the more you'll win. While there are several ways to plan, produce, and present your listing presentation, the basics will always apply.

Foremost, you must know your market. Feed your mind with empirical real estate data. Keep abreast of the circumstances that affect the industry and the economy. Staying informed with factual real estate data will not only make you an exceptional real estate agent but will also position you as the perfect candidate to price properties to sell. Besides gaining real estate knowledge, practicing and knowing how to use your scripts to handle objections is another part of the preparation process.

Preparing for a real estate listing appointment is by far, one of the most vital aspects of your real estate business. It would be best if you concocted a plan to close the deal. The time and energy you invest in preparing for each presentation will determine whether the seller will list with you.

Deliver the Pre-Listing Presentation Packet

A pre-listing presentation packet prepares the sellers by telling them about what will happen at the listing appointment. Besides introducing you and your real estate firm, the package gives the sellers a general idea on how you will help them.

Hand-deliver your pre-listing packet to the sellers' house along with a bouquet. Make sure it includes a handwritten note with your contact information on it. Otherwise, send the package via postal or electronic mail while you prepare for the appointment. Then, send a video email message welcoming them to call you with questions.

The primary purpose of the pre-listing packet is to test the motivation and seriousness of the sellers. When you go to the listing appointment and instantly find out that The Seller's Homework Packet is complete, with two spare house keys ready; you know you got the listing—or at least, they're serious about selling their property. When sellers complete all questionnaire forms, you'll also save time as you can quickly review completed information with them.

Here are some pre-listing presentation elements you can use to send your potential sellers:

1. A Cover Page
2. A Thank you Letter
3. Top Reasons to List with Me
   a. Biography
   b. Clients Testimonials
   c. Printed Online Reviews
   d. Social Proof
   e. Marketing Plan
4. The Top Reasons Homes Don't Sell
5. The Sellers Homework Packet:
   a. Property Information Questionnaire: Why is your home special? (i.e., home features we enjoy, the person we think would love our house, how do you describe your house to a buyer, and what are the elements you like best about your neighborhood? Tip: Use the seller's answers for the AD COPY).
   b. Home Upgrade Form: Interior and Exterior. Also, Cost vs. Value.
   c. Property Information Form: For listing on the MLS, and to make certain information online is accurate and matches the seller's provided information.
6. Home Improvement Tips: The Top Tips on Preparing Your House for Sale
7. Home Staging and Showing Tips: The Top Tips on Staging and Showing Your Home
8. Agent Interview Sheet: Top 10 Questions Sellers Must Ask Agents Before Listing Their Home

Add what you feel is necessary, but do not include documents in connection with the price of the house. Order large envelopes from companies like *ULine.com* and sign up for discounted stamps from *Stamps.com*! This way, you can mail pre-listing packets to sellers before meeting with them. Also, partner with a local florist shop to get a discounted price on all orders. A pre-listing package isn't complete without a bouquet!

> **Take the Next Step**
> *Execute Tactic 12.0: Create Your Pre-Listing Presentation Packet*

Prepare a Comparative Market Analysis Report

Numbers don't lie! Prepare a Comparative Market Analysis (CMA) report to estimate the home value of your sellers. Platforms like *RPR, HouseCanary,* and *Cloud CMS* are a good starting point for CMA reports. Use the most recent and most relevant listings and sales you can find. Pick comparables close to the seller's location, and those that match the house age, size, and style. Analyze a maximum of three months' worth of historical data. In conjunction with your analysis of the recent sales history of similar properties, consider the current real estate market condition as it is also crucial in determining the listing price. Rely on data to conclude what you believe the actual market value is for the seller's house.

Don't just print the latest report on comparables and take it to your presentation. That material is not only limited, but it's often ambivalent. It does not dissect and examine each comparable. For example, drive-by "Comp A" to find out why it took several months to sell and at a lower price than average. Is it in a noisy neighborhood? Why did "Comp B" sell at a higher price? Did it have upgrades that increased its value? Did "Comp C" sell faster than usual because sellers were desperate and willing to accept an offer below market price? Does your research show that "Comp D" took a long time to sell because it needed proper advertising? Did "Comp E" sell quickly and at a much lower price? Did the previous homeowners receive a Notice of Default (NOD) and were about to Foreclose? How about "Comp F" listed by an out-of-territory agent who wasn't even familiar with the town and its neighborhoods? Setting a price using only the CMA report without analyzing the pricing strategy behind each comparable is detrimental and does your client a real disservice.

Keep in mind, to come up with price options that guarantee a sale, you must also consider other aspects that make up a faultless home value. Understand the characteristics and architecture of your client's home—grasp on the potential homeowner who would buy in that specific neighborhood. Think like an appraiser—from touring the area, the seller's house, to the surrounding communities. Consider upgrades made to the home or repairs needed. Know the adjustments an appraiser would make to the comps used. Leverage this opportunity to highlight your expertise by explaining why and how you are making the adjustments.

Moreover, research market climates, absorption rates, interest rates, pending sales, and inventory levels! These factors and many more can affect pricing. It's essential that you comprehend them and then advise your clients properly. Edify yourself and your clients, plug into what's happening locally and nationally, and how it all may affect sellers and buyers. Muster all indispensable ingredients that contribute to making reliable price estimations. Don't just print the price options on a piece of paper and hand it over to the sellers. Scrutinizing is fundamental and a necessity.

Once you have a concrete idea of how to price the house, create a seller's net sheet estimating costs and commissions. This way, you get to the bottom line of what matters to the sellers—that is, the net value. Consider preparing three value options! The lowest listing price you feel would attract the most offers, a fair market value, and the highest value for which the property would realistically sell. When you are ready to present those options, allow the sellers to self-discover the price based on practical data and conviction instead of emotions.

Present Yourself

To convince the sellers you're the right agent for the job, you must first professionally present yourself. You are about to walk into a stranger's house. First impressions matter and are as important as mastering your listing presentation. Show up physically and mentally prepared. Dress well and show up on time. Respect your client's time, stay organized, and put your cell phone away. Acknowledge family, friends, children, and pets to create a rapport with the seller as soon as you come in.

Further, mentally prepare by knowing your craft and your clients. Do your homework and find out why the owners are selling their home. Learn about the condition of the house, and what can improve the residence to sell fast and at the highest price. It would help to pay attention to their financial situation and get to know them on a personal level. Look them up on social media to get a general idea about the potential residents. Knowing your clients is imperative in building rapport and positioning yourself above the average agent.

Moreover, don't underestimate how small characteristics or behaviors make a big difference. For example, showing up to the appointment fifteen minutes early may give you an advantage over the competition and earn the opportunity to sell the house. Also, be considerate of sellers' expectations. The rule of thumb is not to underdress or overdress. Sellers will expect a professional, yet wearing a full suit may not always be appropriate. Dress according to your personality and the area where sellers live. Be sharp, clean, and know that people will analyze and size you up.

Hook Your Sellers

A mediocre real estate agent walks into the seller's house, has a small talk, opens the listing presentation, and reads off the slides. That is NOT YOU. If you want to land a listing, you need to be unique. You have learned the art of storytelling in chapter 4: ELEVATE THE POWER OF YOU, and the elements of persuasion in chapter 5: SPEAK YOUR MESSAGE WITH INFLUENTIAL CONTENT, not to mention the power of pattern interrupts. Combining this knowledge to impress your sellers will set you apart from the herd. Use compelling stories to hook your sellers and get their attention. For example:

"Last week, I closed a house in your neighborhood 20% faster than the average agent and at 6% higher than the average sales price. Would you like to know how I did it?"

Now, you have the sellers' attention! Walk them through the closed example. Show them the result first, and then how they will happily get there, going through the real estate process, and back to the listing presentation. Here's another example:

"I have worked with sellers in different areas and with different home values. I've learned one fundamental fact they all have in common; would you like to guess what that is?"

After asking this question, the sellers' suggested guesses signify the most cardinal concerns they have. Then, you can use emotional intelligence to appropriately and adequately speak to their needs.

Elevate your values and change sellers' perception of you as an agent. Getting sellers' attention and keeping conversational dialog will make your presentation relevant and valuable. Stand out from the competition by addressing what matters most to them first. Hook them with enticing little stories and intriguing questions.

Have a Seller's Mindset

There needs to be a significant change in mindset about listing appointments. Agents with a fixed mindset that think the purpose of a listing appointment is to win the listing is the reason they walk out of the seller's house without the listing. While most agents think the goal of a listing presentation is to get the listing, sellers feel differently. Clients use the listing appointment to determine whether the real estate agent is the hero that will sell their house at the best price and in the fastest time possible.

Therefore, the listing appointment is not to win the listing but aims to identify the clients' problems efficiently. Then, convey to the sellers you are the agent who will provide them the best opportunity to solve their problems and sell the house. Hence, getting the listing happens if you address the seller's challenges very well. Step out of your shoes and think from the seller's perspective. The next time you're driving to a listing appointment, think about the problems your seller may have. It would help to have a clear idea of how to solve them to sell the house, not what you can do or say that will get you the listing.

We can all agree that you and the sellers ultimately want the house to sell. By having the right mindset similar to the seller's point of view, you are now on the same page and within the same frame of mind as the seller. Therefore, you've just accomplished the first step to a successful listing presentation—having a seller's mindset.

The 5 A's to the Agreement Stage

You've aligned yourself with the seller's mindset that you need to reach the agreement stage. Sellers' acceptance of you as their agent will not occur unless you and the sellers agree on every aspect of their needs. Without learning about the sellers' concerns, you neither can solve them, nor will you be their agent of choice. Hence, you won't win the listing unless you address your sellers' needs.

Therefore, the next goal is to determine the problems and then acknowledge them to solve them. It is imperative for you to ask about every need your potential customers may have. While you may not agree with every presented problem, you still have to acknowledge that you understand your clients. Then, address their concerns to build the seller's confidence in you as an agent. Come to a joint agreement on the problems and anything else that stands in the way of the Acceptance Stage—that is, signing the contract.

Build rapport, trust, and take control of the conversation. Follow *The Five A's Concept*: *Ask, Acknowledge, Analyze, Address,* and *Agree* (AAAAA). There are always issues, problems, conflicts, and challenges that a potential home seller may have and that you need to address. Don't be afraid to lay them all out on the table. Ask the sellers about each negative thought that is going through their minds. You'll learn about their concerns once you get them to vent. The customer is always right! Acknowledge the problems regardless of what they are and whether you agree. You build rapport when you focus, listen attentively, and acknowledge their issues. Analyze their needs, wants, and desires. Address every concern with data, logic, and emotions to build trust. When you follow *"The Five A's Concept"* and address all matters, you and the sellers will come into a fair agreement, and you will eventually earn the listing.

You are a real estate hero if you are a proactive, *"5-A's"* type of professional. Don't be the agent on the other side of the table that's playing the defense game. Be the one who dares to extract sellers' fears, faces them, and addresses them. The one that turns evil to good and negatives into positives. Control the conversation and come out on top with a solution to every problem.

The Listing Price Strategy

Since the seller's problem is mostly about price, your listing presentation should focus on price. This way, you will get a sale, rather than a listing. To give sellers an accurate assessment of their home value, you need to arrive armed with comprehensive supporting information, such as a detailed CMA Report. Show the sellers how to create a winning pricing strategy by dissecting the information you have prepared before the listing appointment. Examine each comparable and decipher the reason behind each price. Clarify why and how various real estate data and market conditions affect the listing price. This way, sellers can ratiocinate and arrive at the most logical and reasonable price to list themselves.

Pinpointing a rigorous price and strategically coming up with the fair market value is critical to a successful sale. It would help if you expressed to the sellers that precision is vital. An overpriced home will get less traction, fewer showings, and it will take longer to sell as it demands several price reductions. Sadly, what we convey is the fact that most overpriced homes eventually sell for less than those that are strategically and competitively priced. Thus, don't risk leaving thousands of dollars on the table by overpricing or underpricing your seller's home.

While it is essential to manage the seller's expectation of selling the house at the highest price, be direct and assertive. Be honest and have the conviction in your skills as an agent. Though you want the listing, don't make the mistake of inflating the price so that you can get a signed contract. If you sense that the house won't sell for the desired amount and as fast as sellers wish, be forthright and tell them the truth— even if it means you risk losing the opportunity. By representing the sellers with an overpriced listing, you would risk setting yourself up for failure, frustrations, and potentially disappointing the client. Truthfully interpret the market and have the integrity to tell customers the truth. *"Sally, your house won't sell for what you desire. We need to lower the price!"*

Use both data and emotions to handle the listing price strategy. Your goal is to make sellers realize the right price for the property themselves. Empathize with your sellers, be positive, and educate them on why you feel they should adjust the price. Discuss the reasons and the supporting facts behind your findings. For instance, is the housing market a buyer or seller market? It is acceptable to have a slightly higher listing price if the market is in favor of the seller (or it being in a seller's market). Are there limited mortgage programs and lending restrictions? Are mortgage rates spiking up? If facts show a lack of potential buyers, having a slightly lower listing price might be the right move to attract the most qualified buyers.

Sellers must understand that the price of the home doesn't solely depend on their opinion and their investments in the upgrades. They must take into consideration the market condition. Also, all the other influencing factors, and whether people will be less likely, or more likely to make a significant financial responsibility, such as buying and moving into their house. The pricing strategy of every listing comes down to supply and demand. You want to show potential sellers you care about them by enlightening them with the information that leads to a successful sale. Agree on a price before moving on! There is no point in continuing if you and the sellers do not agree on a listing price.

....
: **Take the Next Step**
: *Execute Tactic 12.1: Create Your Listing Presentation Template*
....

You Are the Agent of Choice

Most agents are often afraid to ask, and sellers don't want to admit. You realize that most sellers are anxious, stressed out, and worried. They are troubled, doubtful, and wondering if they are making the right decisions about whether you are the right agent for the job. Sometimes, sellers aren't even sure about selling their house. Unless you assess the situation and positively change their perspective on every negative aspect that goes through their minds, how else would you connect with them? Have the courage to ask, *"How do you feel about me representing you?"* If sellers are still unsure, address their emotions and the reasons behind their hesitation.

Delivering on your sellers' expectations is vital. Sellers see you as their guide. They want to make sure they hire a real estate professional that is an expert in helping them through their worries. They want a professional that delivers on their needs—selling the house for the most money and within the necessary time. Your job is to demonstrate that you are competent and skilled to do so. It would help if you assured your sellers you are doing what is best for them, and that you will always do what is in their best interest. You want to make sure you convey that you are the backbone to support them throughout the transaction from beginning to end and beyond closing.

Change the seller's overall outlook by substantiating that you are the right agent for the job. Brevity is crucial to success in this arena. Sellers do not want to listen to agents brag about how great they are at selling homes. Before you spend an hour talking about yourself and your company, take a minute to ask the sellers specific questions to find out what services they are looking for from their agent. Find out what type of agent they are looking for to sell their home and then prove to them that it is you. Most people will say, *"We want someone who can sell our house."* In this situation, you have the perfect opportunity to show your confidence. Make eye contact, tell them your track record of success, and ask them if they are looking for an agent of your caliber. If you have a little of a track record, sell your company's history. You may even need to sell a bit of both. Social proof is also substantial. Print and show online reviews or hold your phone up and show them a video testimonial from a past client.

Use Trial Closes and Tie-Downs

If they resist signing the contract, keep digging, continue your script, use trial closes, and tie-downs to gain agreements. One great question to ask throughout your presentation is, *"Do you have questions so far?"* It is one of the best questions to ask, and it's also the one that agents rarely use. The questions sellers ask, and how they reveal what they're thinking would amaze you. Hence, helping you get near closing.

When your presentation is over, always ask, *"What haven't I covered yet that is important to you?"* Listen attentively to see if there are other concerns you've missed. Use other trial closure questions to help you. For example, *"Do you want a lock-box or appointments only? Are you OK* with having an Open House on the weekends? Are there any times that would be inconvenient to show the home?" If you have a concern about the condition of the property, ask to see if the sellers could fix such items. If the house needs staging, also ask if they would like for you to refer them to a professional stager. There are several ways to ask trial closing questions; use a few to test the water and make progress. Most sellers will answer them and proceed forward.

To gain control, increase confidence by getting the client's feedback after each presented idea. During your appointment, use tie-downs to determine where your clients are mentally and emotionally. Are they in agreement with you, or is there further elaboration needed? Are you sensing a "yes" momentum? Are they ripe and ready to be closed? Ask tie-down questions during your sales presentation. For example, *"I can't think of a better way, can you? Who wouldn't want that? Right? Do you see what I mean? Does that make sense? Does this sound good to you? Do you agree with this? Are you with me on this? Great solution! Right?"* Although using trial closes and tie-downs is super important in the listing presentation and sales process, please don't overdo it! The best thing you can do is practice them to use them naturally when you feel they're necessary.

The Acceptance Stage

When you have set up a few trial-closes, and you've agreed on the price, you have arrived! You have arrived at the moment of truth to ask for the order. It does not have to be elaborate, just ask. Here are a few examples: *"I think I have all the information I need; I need your signature in the box, please!"* or *"Can we sign now?"* or *"Do you strongly believe that I can sell your house?"* When they say yes, hand over the pen right away. If they say no, use *The Five A's Concept* to dig deeper. Ask, *"Can you please tell me why?"* Listen to the answer, and then handle their concerns. Again, ask them to sign. Do not give up after the first setback. You will usually breakthrough after the sixth client refusal. You need to be persistent and not give up. If you firmly believe that you are the agent of choice, that belief will eventually come through. Just keep on peeling and asking more questions. Sellers want to pick winners to sell their home.

Many agents do not understand the concept of brevity. A dynamic listing appointment should be short and informative; not a few hours long. Sellers want to know how you can solve their problem, and whether you are the best agent for the job. They don't care about the rest of the presentation. Spending hours on features or other aspects will break your momentum, weaken your position, and your listing presentation. Stay focused, be direct, and get to the point. If you want your clients to choose you as their agent, focus on the problem and the solution, then get the contract signed. That is it!

Finally, once the sellers sign the agreement, spend a few minutes debriefing them. If you have a supporting team, introduce them and their functions to the sellers. Also, fill them in on your routine of communication and the next steps. Taking a few minutes to set the right expectations and explain your system will save you the frustrated seller phone call in a few weeks. Let them know you care, appreciate the opportunity, and move on to the next appointment.

The Home Improvement Stage

On your first tour, give sellers your recommendations of work that will help the house show better. It may be as simple as having the house thoroughly cleaned, the windows washed, the carpets professionally shampooed, the trees trimmed, and the lawn mowed. Your suggestions might be more significant and as extensive as removing a wall, updating kitchen cabinets, adding a backsplash, or installing a new floor. If sellers have had ideas about pre-sale work needed on the house, discuss that as well. Also, talk about which improvements might be worth doing and why. Don't forget to take into consideration market conditions and their expectations from the sale.

While most agents prefer to make work recommendations after clients sign the contract, I recommend otherwise. There are several mutual benefits for discussing potential home improvements on your first tour meeting with the sellers. If you followed along, you do the tour, go back to your office to research and prepare the CMA report, and then set up the listing presentation appointment. If you advised the sellers to replace the carpet, and on your second interview, you had to take your shoes off, then it might be your lucky day. How would you feel when you come in and see the sellers have met your recommendations? It shows not only that the client is sincerely earnest but also that you are already making noticeable progress. Besides, giving sellers ample time to get quotes on all improvements will help them in deciding what upgrades they will be conducting. Also, factoring in work expenses contributes to figuring out the price point of the house.

Don't wait until they've signed the listing agreement to have the maintenance and upgrades discussion. You and the sellers might end up disagreeing on the price and potentially losing the deal altogether. Sellers must have all the facts upfront, rather than finding out three months into the listing what they could have done in the beginning to help sell their house faster and for more money.

To help your clients with any improvements they want to make, recommend your trusted partners. Don't just refer them to the cheapest contractors in town, but only the ones who do excellent work. Keep a list of contractors and vendors who have shown themselves to be reliable, efficient, trustworthy, and cost-effective.

Take the Next Step

Execute Tactic 12.2: Create Your List of Trusted Vendors and Affiliates

Staging the House

The average agent might give clients tips on how to make their homes more attractive and stop there. However, how does that help in marketing and selling the property compared to professional staging? A survey by the National Associations of Realtors® (NAR) said that professional home staging could increase a home's sale price by 6% and sell approximately 17% faster. See the report by searching Google for *"2019 profile of home staging by NAR"*. Printing this report and putting such statistics in front of your sellers will increase the chances of considering a professional staging.

The National Association of Exclusive Buyer Agents (NAEBA) also surveyed brokers and agents. The professionals surveyed said that 82% of home buyers are less focused on any negatives in the home when staged well. These buyers not only fall in love with the house, but they will pay more, sometimes even overpay to get the place. You can hide nothing, and you must fully disclose all issues the house has. Nonetheless, professionally staging the seller's home takes the emphasis off contrary appearances and reveals its positives, leading to more offers and a quicker sale.

Even if Sally has an instinct for decorating, it doesn't mean she understands staging. Offer your clients a staging gift in the form of a no-obligation *"Free Staging Consultation."* Assuming you already found a staging partner, you could invest as low as $20 and as high as $50 per hour. A few hours on a professional staging consultation will work wonders for you and your clients. Such little investments and small contributions make a huge difference in winning the sale. Besides, once you build solid relationships with your partners, many stagging vendors in your network will gladly provide your potential sellers with free staging consultations.

Schedule your stager to meet with the seller within 24 hours of signing the listing agreement. While most agents don't attend those meetings, I highly recommend you do so. This way, you can learn what the "Stager" does to make the house show better. Show up early and introduce both parties. When you're there, don't intervene! Let the professional stager make the recommendations. However, you're not showing up to learn how to become a stager, but to gain more experience and credibility in making improvement recommendations to future clients. Having staging experience is another advantage that gives you an edge and sets you apart from the competition. When a professional stager makes similar recommendations to what you had advised sellers during the first tour, it will help reaffirm suggestions to clients and show that you're multi-talented.

If the seller opts out of the professionally staged home, you and your team might help with that. You can go to local stores such as *HomeGoods, Target,* or *Kirkland's* picking up a variety of different products to use for staging. For example, you can purchase dishes and baskets for the kitchen, throws for the living room, and towels for the bathrooms, pictures, centerpieces, air fresheners, and knick-knacks. Buy neutral things, and light in colors such as white and beige so you can use them anywhere at any house. Your investment in props like these doesn't have to be huge. You would amaze your clients at how much these small efforts add to the attractiveness of their home.

Take a final walk-through after the house is fully staged to make sure everything looks great. Only then should you move to the marketing step and schedule for a professional photo and video shoot.

....

: **Take the Next Step**
: *Execute Tactic 12.3: Compose Your Free Staging Consultation Policy*

....

> *"The only thing that stands between you and success is marketing."*
> —James Tyler

Marketing & Advertising Your Listings

After you do the staging of the house, you can bring in a digital marketing specialist to meet with the client. Your marketing specialist will extract information about the house to use in advertising the property. The marketer will interview the sellers asking questions in connection to what they love about their home, their neighborhood, and the surrounding area. Such information will help draft enticing messages and content for the single property website, flyers, postcards, social media posts, search engine optimization (SEO), display ads, and retargeting ads. This way, you and your team can market the property effectively, and drive thousands of online and offline viewers to the listing.

Send copies of your work to the sellers to express how hard you and your team are working to sell the property. For example, when you mail an EDDM campaign, include your sellers mailing address, so they also receive a copy of your mailer. When you share a post on social media, tag the sellers so they can re-post and share the content with their network. When you advertise the property in the local magazine, hand-deliver a copy of the print to the sellers. Once your photographer produces the photos and videos, and once the content marketing material is in place, email the sellers. The more you get your sellers involved in your marketing strategy, the more respect, trust, and referrals you'll earn.

People usually know many people similar to themselves in income level, interests, likes, dislikes, and perhaps even their profession. Tap into your entire database and your client's network. The people who will be most interested in their home are most likely those people are like the sellers. Tapping their area of influence adds to your pool of potential buyers—and therefore, potential new clients.

When the house is on the market, the average agent might make a phone call once or twice per week to check-in. However, sellers don't want you to "check-in." They are eager and want to hear detailed information about what's going on with their house. Place a lockbox on the property so you can track showings by other agents. Ask for specific property feedback each time they access the house. Then share that feedback with your clients via email—blind copy the sellers to keep them informed when communicating with buyers' agents after each show.

Adhere to the *"Communication Guarantee"* policy and make sure you stay in constant communication with your clients. Talk to them and update them at least twice per week. Tell them how many postcards and flyers you've sent, how many brochures you've hand-delivered, and a list of new postings on the single property website. It would help if you also informed them of how many showings the house has had. When updating your sellers, take advantage of the video updates feature provided by *BombBomb*. Also, include them on your weekly market update emails. Send them text messages and follow up with weekly phone calls to make sure you answer all your clients' questions.

When you are so thorough and precise about all the efforts you and your team are making, you leave no chance for a potential confrontation. If for whatever reason, the house isn't selling as quickly as the sellers had hoped, they know it is not because you are not doing everything possible to sell. If you were only holding an open house or sending out a few flyers, and the house wasn't selling, you might have nightmares. Stay transparent and keep your clients informed by showing them you are doing everything you can take the heat off. It is also how you earn your full commission without discounting.

Enrich Your Listing Presentation with Statistical Data

A big part of being innovative is adopting technologies that can make your business more efficient. It's about understanding how people today are finding homes using information and technology. It's about how and where you market yourself and your clients' properties. Use every new platform and application available to you, implement them in your business, and update them regularly. For example, you can find reports and statistics about how buyers shop for homes for sale. Disclose how they do their research, what devices they use, what buyers find valuable when searching for listed homes, and what tools they deem most useful online.

Here are a few stats from the *"NAR's Home Buyers and Sellers 2018 Generational Trends Report"* regarding the sources buyers used in looking for homes for sale: Websites: 95%, real estate agents: 89%, mobile or tablet search devices: 74%, and online video sites: 39%. Buyers who found their new home by a yard sign or open house was only 7%!

87% of all buyers purchased their home through an agent. 52% of all buyers primarily wanted their agent's help to find the right home to buy. Those buyers were also looking for help to negotiate the terms of sale and to help with price negotiations. 48% of buyers find their real estate agents through referrals.

In a nutshell, if you're still relying on offline methods to market your listings, you need to shift to online advertising immediately! People are searching for homes mainly online and through their mobile devices. They are hiring real estate agents to represent them and negotiate for them. Visit the NAR's website each year to download the most recent report. Then explain these statistics to your clients to illustrate how strategically you market online, targeting potential web and mobile buyers in your territory.

By using these data-driven statistics, you can show sellers how many buyers use their mobile devices for their home search. Then reveal how you use your mobile phone and SMS to forward property-specific website information. You can accomplish this method using the *Text*

Keyword to ShortCode[i] Sign Riders using services such as *AgentText.com* or *SimpleTexting.com.* Also, show city demographics, school scores, and more to prospective buyers. Then show them what a potential buyer sees with your mobile application. Explain how you get notified with each specific buyer's direct cell phone number and how you're able to contact them instantly. It will deepen the seller's confidence in you selling their house and proves to them that you're an expert in both real estate and technology. What else can they ask? The key is to show, not just tell!

Employ High-Quality Marketing Materials

Let me emphasize this: The quality of your marketing material reflects on both the quality of the home and the quality of your real estate services. Typical agents pull out their mobile phones, snap a few shots, and call it good. *"If I am the seller and you do that, I will kick you out."* Their video tours are slide shows of the photos they took with their phones. Their flyers are cheap throwaways that are more self-promotional than designed to feature the house. Please don't do that! If you want to dominate, you need to do better than that. Way better!

After you stage the home to look its best, bring in a professional photographer and videographer. They don't have to be costly, but they must have drones and gimbals. You can find talented college kids or people on *Craigslist.org, GetVermeer.com, Freelancer.com, TourFactory.com,* or even use college students who are training in photography and videography. Unless you have done a lot of photography and video, they'll know more than you about lighting and how to use angles of shots to make something look more impressive.

Even if you don't have the money, don't just use pictures and string them all together. If you must shoot your video with your phone, then spend $20 to have a video editor from *Freelancer.com* professionally edit it for you. Just know of and abide by *Article 12 of the Code of Ethics[ii].* Do not alter listing photos to where they are misleading. For example, a designer can brighten up the sky color, but a designer can't use Photoshop to change the color of a wall or remove overhead power lines. Keep in mind, you can do anything less expensively and still do it well.

To produce even better quality digital assets, use a drone for footage if you have access to drone photographers. Nothing can fully feature

---

i   *See "Text Messaging" under the "Marketing Automation" section in chapter 6: SHARPEN YOUR AXE.*

ii   *https://www.nar.realtor/sites/default/files/policies/2007/code-of-ethics-article-12-2007-11-02.pdf* the property more than drones and balances the stability of videos like gimbals. Also, an SLR camera to take the highest quality photos is a MUST. Use the drone shot as the first picture on MLS and everywhere on the web, including your marketing website. It will draw attention to the house and you as an agent. Remember, marketing is about being unique and going above and beyond the customer's expectations.

If you want high-quality marketing materials, spend more than the average agent spends. Invest in yourself by investing in the quality of your business and your clients' listings. However, your budget doesn't have to break the bank. When implementing some of these ideas or concepts you think of, stop, and think about the most affordable ways to get the job done before you even start. As your revenue increases, add to your budget and expand your business. Implement, improve, grow, and repeat! Always focus on getting better in quality, technology, and automation, so you can improve your marketing and land more deals.

Expand to Social Media and the Internet

For each video you produce—whether it's an educational or a listing video—promote it to your social media pages. If you haven't done so, upload your 6,000 contacts gathered in previous chapters to *Facebook Custom Audience* and *Google Custom Audience*. Advertise to them on Facebook, Instagram, Google, and YouTube—including Display Advertisements and Networks.

Follow the *"Hook, Story, Offer"* model to attract potential buyers to your listing. Create a *hook* using pattern interrupt to catch your audience's attention. Tell a short *storyline* that gets people interested in wanting to find out more about your listing. Then the *offer* is the value that captures leads. The offered lead magnet can be an eBook download or a CMA report. With a $400 monthly marketing budget invested into Facebook and Instagram ad campaigns, and to your 6,000 people in your territory, or Geolocation ads using Facebook Map, you can generate, on an average of about 60 leads using a 10% overall conversion ratio, which is on the low end. Here's the breakdown of this scenario:

- 6,000 contacts x 10% conversion ratio = 600 Click-Through Rate (CTR)
- 600 clicks or visits to landing page x 10% = 60 potential leads

Moreover, anyone who viewed your videos or visited your website has been *Cookied*. They all become a part of your retargeted audience. You will show up on their feeds, stories, sites, computer, and on their mobile devices. Anywhere they go online; you're there! That is the power of omnipresence.

Therefore, when presenting to potential sellers, you should include all these online and social media tactics in your Digital Marketing Plan. Visit chapter 6: SHARPEN YOUR AXE, to learn about pixels, ads, retargeting, and all the tools necessary to accomplish a successful marketing strategy. Use companies like AdWerx or PerfectAudience to expand your retargeting advertisements to social media and the web.

Engage the Real Estate Community

If you are a new agent, engaging the real estate community through brokers' tours is an excellent method for getting to know other local agents and brokers. Brokers' tours are great for meeting new real estate professionals, developing rapport, and examining your competition. Getting to know other agents in your community, and especially top ones, is very important. When deciding which buyer's offer to choose, often, sellers will ask your thoughts about other agents, and if you had any experience with them. Hence, connecting with your peers is helpful to you and your future clients.

Furthermore, don't rely only on the broker's tours to engage your real estate community. Email *"Coming Soon"* PDF flyers to inform your local fellow agents throughout the county about your upcoming new listing. Those agents can then forward your flyer to their clients in case they are interested in making offers on your listing. They will thank you for potentially generating them buyers, and also for making them look great in front of their clients using your flyers.

Additionally, some sellers think their property would sell much faster if they hire a real estate professional who works for a prominent real estate firm with a large number of agents. They want to leverage the benefit of the vast reach such an agent may have. You can overcome this objection by using *Proxio*. *Proxio* allows you to market your listings to over 650,000 agents in 15 countries and 19 languages. If you don't have a listing, speak to a few agents in your local area to see if you can use their listings to generate your leads. Most agents will be delighted to accept your offer. This strategy is a win-win situation for the both of you. Check out *proxio.com* to send out digital flyers and to tap into all agents in your community. That is how top producers at large real estate companies connect.

Similarly, you can even take this tactic a step further and email all the top agents within the three to five surrounding counties. When the market is slow, or it is a buyers' market, it becomes hard to separate your listing. Mail property flyers and brochures to all real estate offices within your territory, especially when the market inventory is high, and you need your listings to stand out. Also, attempt to meet with top producing agents in the area.

On the contrary, avoid the last two tactics in a regular market, or when the home inventory is low as they are unnecessary and time-consuming. Thus, they make the most sense in a buyers' market. However, in a sellers' market, promoting your listings at Brokers Preview events should be sufficient. Therefore, be productive, engage your community when needed, and do what is in your clients' best interest. Leverage other real estate agents, your farm, and your neighborhood wisely.

Elevate Values and Exposure

Use your marketing strategies in advertising your listing for maximum exposure; this includes marketing efforts from MLS, Sign, Lockbox, Flyers, Ads, Internet Presence, Social Media Networks, Open House, and more. Without a doubt, the solid marketing plan and all the guarantees created in chapter 3: DISCOVER YOUR CORE VALUES will set you and your real estate office apart from the competition. Your goal is to elevate your values so high that no other agent can even get close to the benefits you bring your clients. Ask the seller, *"Do you know a real estate professional that is doing even half that much for every listing?"* The expected answer should always be a firm, "No." If they say, "Yes." Ask if they can share the name of that agent. That is your competition!

Top agents exhibit their competence by always elevating the values they bring their clients. Increasing the perceived benefits warrants them to charge higher commissions than anybody else. That is because they give more value than anyone else. They also prove it through expertise, marketing, and social proof. Suppose the standard commission is about 4½% to 5%, while most agents might charge less, top agents will charge 6%. Be confident in your capabilities and earn what you deserve through your marketing guarantees and strategies.

If it is a seller's market and you are the listing agent, you can charge 5½%. Take 3% and give 2½% to the buyer's broker. If you are in a buyer's market, consider charging 6%. This way, you would incentivize the buyer's broker with 3%. You would also take 3% and still provide your clients with the same outstanding services. In a buyer's market, offering a higher commission to the buyer's agent will ensure that the listing gets shown more. You and your sellers will end up with quicker sales. Pay more and sell faster; you will not only create happier clients but will also equate to more money and more listings.

Therefore, when your clients see how much more you provide compared to the average agent, they will not complain about your commission, nor will they resist. People should come to you because they recognize the difference and the vast gap in value perception between what you do and what others do.

To outweigh the 3% commission, you must exceed expectations to show clients the benefits you provide. Break down the money you will invest by making sure they end up with the best results. The more you do that, the more your customers will brag about you and appreciate your services. When you get an objection, pull the net sheet and show them the bottom line of what you'll get them. Don't be the agent who will charge them a low flat fee to win the listing! Be the agent who is willing to negotiate the best deal, and promote the listing the way you would market yours. Hold steady and stick to your guns. You got this!

When you go the extra mile for your clients and keep your values high, neighbors see the same benefits, and they notice that you're different. Like I mentioned throughout the entire book, you introduced yourself, they know who you are, and you prospected them before—or did you? Also, you dropped off your high-quality brochures at their houses. Right? They see you plastered on social media, marketing other homes in their neighborhood, and sent voicemail drop campaigns. Besides, you paid for promotional ad campaigns on YouTube, Facebook, and Instagram. Your ads are inescapable, and your omnipresence is unavoidable! Again, these marketing tactics explained throughout the entire book was to get you here. When it's time for those neighbors to sell, they'll automatically think of you, message you on social media, email you, and call you. You want to thrill your clients with your high-quality real estate services. You need to be extraordinary! Clients would want to brag and boast about you before they even have an offer.

In close, use the below 5 E's model as a roadmap to prepare the best listing presentations. Explain your marketing plan and sell the property at the best price and in the fastest time possible.

1. **Enrich**: Enrich Your Listing Presentation with Statistical Data
2. **Employ**: Employ High-Quality Marketing Materials
3. **Expand**: Expand to Social Media and the Internet
4. **Engage**: Engage the Real Estate Community
5. **Elevate**: Elevate Values and Exposure

Then, use the acronym MAPCAVES in your listing presentation to help you walk through the closing steps:

- **Mindset**: Whose mindset should an agent have at the listing presentation? The seller!
- **Agreement**: Whose problems and needs should an agent agree? The seller!
- **Price**: Who determines the listing price of the property? The seller!
- **Condition**: Who controls the status and the state of the property? The seller!
- **Access**: Who allows selling agent the ability to get easy access to the property? The seller!
- **Value**: Who determines the value of the property? The market.
- **Exposure**: Who controls the exposure and the marketing of the property? The listing agent!
- **Staging**: Who decides whether to stage the house professionally? The seller!

> *Take the Next Step*
> *Execute Tactic 12.4: Engage Your Community*

Escrow Services

Inform your sellers of the escrow services and explain exactly how the process works. Give them access to your Transaction Coordinator (TC) in case they have questions regarding the sales process. Sellers must be abreast with every step of the transaction from opening escrow, to delivering disclosures, scheduling inspections, handling repair requests, if any, and to close.

As the listing agent, always get the seller's permission to meet with the termite inspector, workers for repairs, or to install signs and lockboxes. Even if it is in the contract, don't just show up unannounced. Also, oversee the entire process and provide all disclosures. When the appraiser comes by, attend and bring your CMA report with all upgrades, and multiple offers, if any. Doing so will help appraise the property at the right price that the house deserves.

During this final stage, it's especially important to make sure your client understands what is happening, primarily through the escrow process. Sellers often think when they go to sign the closing documents, they'll be handing over their keys and picking up their check. Don't let your customers get super excited—then very disappointed. Make sure they understand the escrow process, especially the timing of this last piece, by sending out another *BombBomb* video after the offer acceptance stage that explains it all. However, always be there in person for the final walk-through and to deliver the keys to the selling agent.

"Winning isn't everything; it's the only thing."
—Vince Lombardi

Negotiations and the Offer

I recommend negotiating all offers in-person because of the effectiveness and the efficiency of everyone being physically present— sellers, buyers, and agents.

If you're the listing agent and an offer comes in, set up the presentation at the seller's house. The selling agent must present it directly to the sellers and you, the listing agent. Buyers should wait outside reviewing disclosures while waiting to respond to counter. This way, you complete the negotiation process in one meeting. Otherwise, you may prolong the negotiations process by accepting offers via *DocuSign*, Email, or PDF.

If the listing agent accepts offers electronically, ask your buyers to record a *BombBomb* introduction video to send the sellers along with the offer. For example, *"Hi Sally, we love your house, and we would love to buy it. We are sending you an offer of X dollars. We are hoping you'll accept our offer so we can get this process started. We have a loan approval from our lender at the offer amount we presented. We are ready to proceed as soon as you accept our offer. Thank you for your cooperation. Bob and Barbara."* A pre-recorded personal video, along with the offer, can work wonders.

Before Writing the Offer

Explain the fine print in the contracts, forms, and addendums. Sellers don't understand most of what's on all the paperwork. It is your job to walk them through every item and ensure they fully understand the consequences of every responsibility. Leave nothing out!

Advise your clients on how to disclose appropriately. Give them examples of what a proper disclosure looks like and add an addendum to the contract to elaborate on any changes they've made to their home.

For example, on the California Association of Realtors® transfer disclosure statement, one question is, *"Has anything ever been repaired, replaced, modified, added, fixed or changed in the home?"* It is unlikely for sellers that have lived in a house of many years, not to have fixed a broken faucet, a water heater, garbage disposal, or a garage door. Disclose even a simple installation of new light fixtures. When, in fact, most disclosures usually say "No" to this question, the agent should coach sellers to disclose any repairs or modifications done to the house!

Qualifying Buyers

If you're working with a potential buyer and qualifying is in question, don't just provide contingency protections for the sake of accepting an offer. As the listing agent, you must pro-actively qualify buyers for a smooth close. You are probably not the average listing agent who gets an offer in and presents it based on whatever information the buyer's agent has given. Right? Most average agents add contingencies to the counteroffer, giving the buyer time to qualify for a loan. Don't do that! Only work with qualified buyers who have proven affordability, whether cash buyers or those with official loan approvals from their lenders. Therefore, before you accept an offer, make sure the buyer has been truly qualified. Don't waste the seller's time with a buyer who may or may not qualify. Verify prequalification by having the following items along with the offer:

1. A Conditional Loan Approval letter from the lender.
2. A confirmation letter from the lender with verified employment, income, and assets.
3. A copy of the tri-merge credit report showing buyers' three credit scores.
4. Proof of Funds (POF) such as a copy of a bank statement or letter showing sufficient money to complete the transaction.
5. The offer.

Please note, if cash buyers, skip items 1, 2, and 3, and only look for the POF along with the offer. Otherwise, it would help if you find out how long the lender has been in business, how active they are, and whether they work in real estate full-time. Also, as the listing agent representing the sellers, avoid any potential losses or failed transactions by spending some time vetting the buyer's agent. This way, you can give your sellers an honest opinion about how the sale would probably go based on their offer.

....
: *Take the Next Step*
: *Execute Tactic 12.5: Create a Customized Home Buyers Questionnaire*
: *Form*
....

After Offer Acceptance

After sellers accept the offer, don't just wait for the buyer's agent to do the heavy lifting. It would be best if you, the listing agent, were in control of the transaction. Provide your clients and their buyers with excellent service. Remind your sellers of their responsibilities and make sure you stay in constant communication with the selling agent. Send reminders the day before every appointment via text and email. Three days before any contingency deadline, send the buyer's agent a reminder of their responsibility and offer help, if and when needed.

This process doesn't have to be tedious and time-consuming. Your CRM can immediately generate reminders once you enter the dates. Automate all reminders throughout the entire process via your CRM platform. This way, you can save time, ensure a smooth close, and a happy client. Another helpful tool you can use to manage your customers and the workflow of every real estate transaction is *Folio by Amitree*.

Working with Buyers

Nowadays, searching for a new home differs from a decade ago. Buyers now search listed homes themselves, online, and from the comfort of their own home. Besides, most buyers typically know about new houses on the market before their agent does. If they are serious about buying, they're using apps, websites, and digital platforms to find potential properties. They even set up alerts for new listings that fit their needs. From browsing mobile apps like *Homesnap* and visiting sites like *Zillow, Trulia*, and the dozens of others like them, buyers can now browse listings, find homes, and narrow their search all on their own—without ever calling in an agent. Technology has forever changed how people shop for homes.

NAR research shows that 75% of seniors use online platforms to find their next home. Millennials believe that they no longer need agents to help them find their dream home. They can do that and find their next home themselves. Therefore, buyers' expectations and the role of a buyer's agent have changed. Agents need to stay connected and know their market so they're able to locate homes that won't show up online.

To be a thriving buyers' agent who adds real significant value, you need to be unique, innovative, informative, and a great negotiator. Your aim should be to provide exceptional services and equip your clients with all the knowledge and resources they need to make a successful purchase. First, make sure you have the answers to the most common questions asked by potential home buyers:

1. Do you work mostly with home buyers or with home sellers?

2. Can we reach you through various communication platforms like email and text messages?
3. Do you have a list of reputable vendors like lenders, home inspectors, and insurance agents for me to consider?
4. Are you a member of the local Multiple Listing Service (MLS)?
5. Are you a member of the National Association of REALTORS®?
6. Do you have information about For Sale by Owner properties?
7. Are you able to provide local real estate market trends, values, and statistics?

Intimate Local Knowledge

Those selling agents who genuinely connect with their communities, and those who possess extensive knowledge about the markets they serve, win. This includes details like the district's history and the stories of its real estate values, demographic trends, property tax rates, school ratings, crime statistics, amenities like recreation areas and parks, local businesses, shops and restaurants, current and planned commercial developments, available police and fire stations, social services, transportation and traffic patterns, and access to healthcare. Picking up a generic market report that covers these topics doesn't make you the expert real estate agent in the local area. You must be a veteran, and familiar with the region as you have been living there for a decade.

Only serve areas you know and with which you are familiar. You do a disservice to your buyers if you try to represent them in a market outside your territory. If you live an hour away, you won't know the subtle differences between neighborhoods. You might miss disclosing things like higher tax assessments in specific districts or counties, potentially longer commute to work due to often jammed traffic on certain streets, high rates of public parking, or the noisy crowd during the Saturday night high school football game.

Besides knowing your market and your community, part of knowing your areas personally is having a close relationship with local listing agents. Develop a good rapport with them through brokers' tours and attending local real estate events. Attempt to get to know your fellow agents. Be kind, and reach out! It will surprise you how friendly, helpful, and resourceful other agents can be, especially if you're new to the area. Also, be prepared to respond and follow up quickly. Regardless, always deal fairly with everyone, and strive to be useful, fun, and easy.

Knowing the market and building positive professional relationships gives you the inside track. Other agents will be more comfortable sharing their success stories, strategies, and vital information with you. Clients will look at your offers more favorably because once their agents know you, they will like you, trust you, and work with you.

Thoroughly Understand Your Clients' Goals

When working with buyers, agents need to understand their clients' goals, needs, lifestyles, and personalities. A buyers' agent needs to go beyond the typical questions of, *"How many bedrooms and baths do you want?"* You need to dig and understand customer's profound motives. For example, find out the place of your clients' work and how they feel about commuting. How do they spend their time at home, and what do they enjoy doing in their free time? Do your clients expect their family size to change? Are they getting married, having kids, having older kids move out, or having a family member move in with them? Also, do they plan on having any pets? How about hobbies and activities? Are they outgoing? Do they entertain at home a lot? Do they like to go for walks on the beach or at the gym? What amenities are most important to them? Do they work at night and sleep during the day? The more you learn about your audience's needs and motives, the more you'll understand and connect with them.

Find out specifically "why" your clients want to buy a new house. How long do they plan on living in the new house? Are they looking for a forever home or an investment they can turn over in a few years? Is the school district important to them? Do they have kids that need to attend better schools? Of all their reasons for buying, which are the most important? By asking all these questions, you not only gain a better understanding of the home that will work best for them, but you also help clarify their goals and desires so they can make the right decisions. Thoroughly understanding your customers and their objectives can set you apart, help you find the perfect home, and deliver the lifestyle they've imagined.

Educate Your Clients About Home Ownership

One of the many incredible values you can bring to your clients is to educate them about homeownership. With the right to "life, liberty, and the pursuit of happiness," owning a home has always been an essential part of the American Dream. However, for people who don't understand the ins and outs of homeownership, owning a home can become quite an ordeal, as we all witnessed the real estate bubble in 2008. The fact that your buyer has owned homes, it doesn't necessarily mean they understand the potential of damage and devastation or the great benefits of homeownership.

Buying a home is a very emotional stage for many people. You want to set your clients up, so they don't let their emotions lead them into bad decisions. Efficiently educate your potential buyers by creating general educational videos to teach them about purchasing a home. Produce videos with topics such as:

- *The pits and falls; pros, and cons of home buying*
- *When is the best time to buy a new home, and why*
- *The top home-buying mistakes and how to avoid them*
- *Tips for making an attractive, winning offer that gets accepted*
- *How would a market crash affect a HELOC?*

Videos like these will save you time, get you quality leads, and your customers can refer to them at any time! If you have a fear of getting on camera, it might help to start by recording podcasts. The goal is to deliver on your client's goals and needs. Keep in mind, what is second nature to you is a confusing muddle of real estate jargon to your audience. Become resourceful so you can educate, connect, and win more clients.

Other subjects you can cover include topics like the impact of unpredictable market volatility on home equity lines of credit. When the market crashed in 2008, most homeowners claimed that they had just lost thousands of dollars on their home. That "cash" they thought they'd lost was never a fact. Homeowners need to understand that pulling cash from their home equity as their own personal Automated Teller Machine (ATM) can sometimes lead to foreclosure. It would help if you enlightened your clients on the fact that, although real estate can be an excellent investment, its value can go down. I am not saying you need to be a financial advisor; however, to be unique and stand above the average, you need to educate your buyers on the mechanics, and the nuts and bolts of homeownership.

Educate Your Clients About Home Affordability

Do your due diligence and make sure your buyers understand the costs of homeownership, including taxes, insurance, homeowner's association dues, utilities, as well as the potential expenses associated with ongoing repairs and maintenance. Just because a family member is gifting most of the down payment and they can afford the mortgage payment doesn't necessarily mean they should buy a home. Educate your clients about home affordability, especially first-time home buyers and move-up buyers who are moving into more expensive homes.

Find out if your buyers want to have an impound account and include their taxes and insurance in their monthly mortgage payments. It's essential to help them factor in all the costs they'll incur in owning a home so they can calculate what they can truly afford. Even where the mortgage lender requires little to no reserves, educate your buyers on the benefits of having a minimum of six months reserves after closing. Emergency cash is especially necessary for those who have no liquid assets, such as marketable securities, mutual funds, stocks, or retirement plans.

Here's a quick checklist of the basic affordability subjects you should discuss with your potential buyers to set the right expectations:

1. **Property Taxes:** In many areas, the sale of a house triggers a property tax reassessment. Research local tax information, and warn your clients of possible increases. Each County's Treasurer-Tax Collector website can show the history of property taxes collected on average over the past three years. This information is available through public records. A potential homeowner can search each property address, and find out how much they'd be paying each year in property taxes, including supplemental assessments. Often, when the property tax bill arrives, it shocks most new homeowners of how much they must pay each year.

2. **Homeowner's Insurance:** This is another annual fee that most homeowners don't account for each year. Again, make sure you discuss the potential added costs to their mortgage payment each year.

3. **Home Upgrades:** It's exciting to move into a new house and put your personal touch on it. However, one "small" change invariably leads to another and more cost. A new backsplash may require new countertops which may require a new kitchen sink and a new faucet. While a new deck may sound easy and inexpensive, but those projects rarely are. Also, repairs that require getting permits will automatically increase property taxes. Counsel your buyers to be cautious about improvements and suggest they delay them until they've settled in their home.

4. **General Maintenance:** Professional property managers usually create a "reserve for replacement" account. It is a fund set up for the maintenance issues that may occur to the property. For example, repairing the roof, re-sealing the driveway, re-staining a deck, replacing doors and windows, or replacement of furnace and air conditioning. There might be a constant flow

of issues that you might need to address. For new homeowners who have been apartment dwellers, this can be a huge shock. Let your clients know that regular and thorough maintenance of their home will directly affect the price they can get when they eventually sell it.

5. **Lawn and Garden:** Help them get an estimate of landscape maintenance charges. Even a basic mow and blow can get costly. If they decide to do it themselves, they'll incur the cost of all the equipment they need.

6. **Unforeseen Accidents:** If the house is on a golf course, your buyers can expect a few golf balls flying through doors, windows, and shutters. If your clients get many home deliveries or have many guests over often, someone, sometime, might drive over their sprinklers or into their garage door. Kids are notorious for throwing things down toilets and putting holes in walls. Also, breaking doorknobs, light fixtures, window blinds, screen doors, and sometimes themselves.

7. **Safety Issues:** Does the new home have a pool, and does it need a fence around it? Do those bathrooms need safety handles? Are the banisters and railings of the stairs sturdy? How about a new security system and motion detectors?

8. **Emergencies**: Is the house in a flood zone area? If so, what do they need to protect the home? Does the power go out frequently because of storms that topple trees? If so, do they need a backup generator—or at least a bunch of flashlights?

9. **Time**: One of the homeownership's most considerable costs is time. If your buyers are "Do It Yourselfers," working on their new home will eat up most of their free time. Even if they hire vendors to do the work, they'll spend hours waiting for those vendors to show up.

On the contrary, you can also point out the financial benefits your clients will receive in terms of tax write-offs to add to their affordability calculations. In some states, homebuyers can write off some interest paid, which can save them several hundreds of thousands of dollars per year. They can either get that money back at the end of the year or save on their taxes. They can also increase their dependents to save that money each month. When you factor this benefit in, your clients might be comfortable that they can afford a pricey home. The worst thing you can do as a buyers' agent is getting your clients into a home they cannot afford. The best thing you can do is to educate them on all elements of home affordability and ownership.

Educate Your Clients About Home Financing

Don't leave home financing up to lenders. While it is the lender's responsibility to explain all available mortgage options to your buyers, you can't just rely on them. Get involved by making sure your clients understand the difference and the consequences of the various mortgage programs available to them.

Invest the time to explain to your buyers the pros and cons of adjustable versus fixed-rate mortgages, reverse mortgages, interest-only, 3, 5, 7, 10, 15, 20, and 30-year terms. What are the consequences and the benefits of each type of loan, and how will it affect your buyers? Make sure they learn about Private Mortgage Insurance (PMI), and impound accounts, if applicable. Mention the benefits of making bi-monthly payments to pay off the home quicker and incur less interest. If you are staying on top of the market, know what interest rates they are likely to get, and which lenders would be the best match for them.

Do your buyers qualify for any special financing? The financing part of home buying is incredibly essential. Double-check their mortgage Note and confirm what they're signing for is what they truly want. The Note explains the mortgage program, the interest rate, the margin, the terms, and other aspects of the mortgage loan, such as adjustment caps that most homeowners do not even understand what they mean.

In the early aughts of the two-thousands, many people opted for interest-only loans. One cause was that they could not afford the fully amortized monthly payments. Another reason was that the mortgage brokers were making so much more money by selling those interest-only home loans. After the interest-only period was over, people were later shocked to find out that they paid nothing into the principal balance. Plus, the massive gap between what they were paying versus the new fully amortized amount. The only individuals who enjoyed those loans were mortgage brokers. They charged excessive points paid by lenders. Why not? By selling high-risk mortgage loans, these banks had highly profitable margins.

For example, do you remember the *"1% Option ARM?"* It was also known as the *"Pick a Payment"* program. Lenders and mortgage brokers sold this product with the three-year prepayment penalty option for higher rebates. The Option ARM was a high-risk program that allowed almost every person in the nation to own a home. At that time, loan consultants rarely explained the terms and the risks involved to homeowners. Hence, it made people believe that they can afford to buy more homes, only to find out a few years later, that this program was one of the main reasons foreclosures were inevitable. Thousands of homeowners could not pay their adjusted mortgage payments, leaving them without homes. How could they afford a fully amortized monthly payment at 6% up from 1%? They lost everything they've ever had because, to a certain degree, mortgage and real estate professionals didn't educate them properly. Had clients read their mortgage Notes, and professionals explained their terms, many homeowners would not have become insolvent. Thus, it is highly critical that you educate your clients about their home financing options.

Be a leader and have a leader's mindset of "people before things." Make sure you align yourself with great banks that have the same high ethical standards you have. Be sure that your clients are clear about their lending options. Would an extra $400 per month in mortgage payments mean they'd have to cut out date nights, weekend family outings, or doing the things they enjoy in life? Let your clients know, "*If extending yourself for your mortgage means sacrificing life enjoyments, then don't do it! Don't get shackled to a mortgage payment you really can't afford.*" It's devastating to see people lose everything they've worked for because of a change in the real estate market!

Besides educating your clients on their mortgage options, help them position themselves to look suitable to lenders by showing them how to improve their credit score. Perhaps suggest paying off certain debts or increasing their down payment. Then help them get pre-qualified with a reliable lender that has a good track record and extensive experience before they enter the market. Let them know even if they haven't found a home yet; they can get a "To Be Determined" (TBD) loan approval. A lender can fully contract a TBD loan approval up to the entire underwriting process contingent on the appraisal report. A conditionally approved TBD loan is much stronger than a pre-approval letter because there would be no financial surprises further in the process, and it speeds up the closing of the sale.

Delineate the Home Buying Process

Give your clients a thorough explanation of everything that happens from submitting an offer to Close of Escrow (COE). I'd suggest creating a series of videos on the different stages of the process with a checklist to keep track of every milestone. Cover every aspect from offers to purchase contracts to contingencies and appraisals. Tell your buyers what to expect from a home inspection or a termite inspection. Help them answer common questions, such as "*What happens if the house doesn't appraise? What inspections do they need, and is it safe to accept the sellers' inspections?"* Answer every question in detail and address every concern buyers may have. Your clients will feel more confident, and you'll end up answering fewer questions along the home buying process.

As you do this, point out the parts of the transaction particular to current market conditions. For example, if you are in a sellers' market, offers must be more attractive to beat out the competition and get accepted. Buyers may need to move quickly on homes that interest them. They might need to do extras to connect with the sellers. For instance, writing a personal letter to the sellers about what they like about the house, or have their financing consultant call the sellers' agent to verify the pre-qualification and the lender's commitment.

As you get into the transaction, give buyers various recommendations for inspectors, vendors, carpet repair, window repair, and anything they need. Provide multiple references for reputable partners and organizations with which you have good relationships.

Conduct Effective Tours for Buyers

Your job during home tours is not just to be the chauffeur and deal with lockboxes. You need to think of yourself as the real estate expert your buyers have hired to help them make a great decision on their new home. Rather than emailing the standard MLS information on homes you'll be touring, create your own buyers' tour packet.

Email potential buyers a checklist of the most significant factors in every home they seek. This way, homebuyers can check off features they love about each house. After seeing several homes, buyers will often forget a few critical aspects of the ones they've toured. Once you are busy, you'll also lose track of which buyers visited which homes. Your buyers will have a hard time remembering every house they see. Help yourself and help them out by making a custom report for each tour. As you tour through each home, use your laptop, tablet, or your mobile phone to add your comments. Take videos, and record your buyers' notes. Of course, request their permission before doing so. Use *Google Keep* or *Evernote* to send the videos, pictures, and notes digitally. You can also print your comments if your clients prefer that.

How about turning all that into a digital tour package? Create a full-colored report that has additional photos and videos of each house where you leave space for your buyers to make additional comments. Include neighborhood information and a Google map that shows the location of each house concerning others they have toured. Look into *Google Drive* or *Folio by Amitree* to accomplish this.

Prepare yourself by checking in with sellers' agents to get some background on why the house is selling, how quickly sellers want to move, and whether they must rent back. Dig to understand what is important to the sellers. Research the specific location of each home: Are there hazardous landfills and waste sites nearby? Is the home near an electrical power station? Is it near a train track or an airport? Make notes of these essential elements so you can inform your clients about all the details of the location while on tour. It's just as important to point out the negatives of a house as its positives. Point out the issues and disclose all information you deem necessary even if it means you won't get a sale.

Also, go the extra mile and scrutinize each property. Although you are not a certified home inspector, don't just rely on the inspector to advise your clients on potential physical problems with the house. Pay attention to any indications of possible issues as you would if you were buying a home for yourself. Signs like concealed unsightly water stains on the ceiling could mean a leak in the plumbing, and foundation cracks may indicate notable structural damage. Add your notes to the *BombBomb* video and send it to your buyers to refresh their memories after the tour. Again, log your recommendations in *Google Keep* or *Evernote* to keep track of toured homes while visiting each property.

Design a Win-Win Offer

Bob and Barbara, the buyers, have found the house they want to buy. Now, it is time for you to create the most compelling offer possible. Ross the Realtor may advise his buyers to add one to two percent of the listing price to the offer, so they have a higher chance of getting their offer accepted. This method is not "the strongest offer." The most substantial offer is the one that gets Bob and Barbara the house on terms that work for them logically. When your buyers are in love with a particular home, your job as an agent is to help them control their emotions and ensure they are making the right decision. You want them to be happy and know that you did what is best for them years down the road.

You need to be creative and have excellent negotiation skills to draft an acceptable offer. You must approach the negotiation table with a strategic system. From knowing what the traditional fees are in that county, to what price points are negotiable. Also, what's customary regarding closing costs? For instance, who pays for title or attorney fees? Who pays for broker fees, lender fees, escrow, home warranty, and inspections? What repairs should you request from the sellers, and what fixtures they must keep in the house? Remember, the market climate affects what is customary. Everything is negotiable depending on the sellers you're working with and what the market is doing. Knowing your industry and the market norms in an area you're farming is essential to create an attractive offer that sellers would be happy to accept.

Get as much seller information as you can to understand what kind of leverage you and your buyers have. Figure out why they are selling. Are there any hardships involved that may give you the upper hand in negotiating a deal? Look at the length of the listing period; has it fallen out of escrow, and if so, why? How many offers are in and how strong are they? Ask the selling agent if they will respond to all or only counter the best offers. Make sure your clients understand the difference and know when it's a competitive market. If it is a seller's market, let them know they must show all the strength they've got from the beginning and not leave anything to chance. You can only negotiate, but you can't force Sally to accept Bob's offer.

Furthermore, show Bob and Barbara how economic indicators, interest rates, real estate cycles, and market conditions, both locally and nationally, play into the offer. For example, in a sellers' market, Bob and Barbara need to know that Sally will get multiple offers. The house will most likely sell at 3% higher than listed. Also, the home may not even appraise for going price. Hence, their offer must be firm if they want the house. Talk about the pros and cons of making an offer on that home. On the other hand, being in a buyers' market versus a sellers' market will drastically impact how you approach the sellers, other agents, buyers, offer, and the overall negotiation process.

Confirm with your buyers that all requests to sellers are final before making the offer. Educate them about the process and the demands they can make. For example, what repairs buyers should require fixing? Also, advise them on what is a fair offer given the market conditions, so they don't offend the sellers. Don't let your buyers go into a contract, then make unreasonable requests that everyone sees as an attempt to renegotiate after the negotiation is over. Cornering the seller with additional demands right before the close of escrow is unfair and delays the entire process.

Plan your offer by letting the seller's agent know how strong and prepared your buyers are. Ask Bob and Barbara to record a video and send it directly to Sally. Also, deliver the listing agent your buyer's letter that mentions a specific aspect they love about the house. Producing a BombBomb video to the seller is probably best. *"Hi Sally, we love your house, especially the view from the master bedroom. We look forward to spending most of our afternoons in the spacious backyard. We also look forward to meeting the awesome neighbors. We hope you'll accept our offer. Thank you. Bob and Barbara."* Such a video and letter work well as they give Sally the sellers a personal connection with your buyers. Many sellers are only concerned with money, but many others also really want their home to go to a suitable buyer. They want to leave their neighbors with a good new neighbor. You never know what kind of seller you will get, so pull out every tool that works in favor of your buyers. Remember, personalization and emotional intelligence are highly effective, throughout every step of the real estate process, especially the offer.

If the transaction isn't a cash offer, it means that your buyers had to get a home loan. To strengthen your client's situation, have the mortgage consultant call the seller's listing agent directly. The loan officer can explain how qualified your clients are and how quickly the lender will close. The consultant should also elaborate on how long the lender has been in business and how great of a relationship they have with you, the real estate professional. This tactic will not only show that the bank can deliver on their promises, but will set you apart as an agent because of your *supercalifragilisticexpialidocious* relationship with the lender. Combined with the personalized video, this third-person feedback will support the emotional connection even further providing additional credibility to both you and your buyers, leading to an excepted offer.

As negotiations continue, explain each part of the agreement to your buyers. Make sure they understand the actual cost of homeownership. For example, explain what it means to them if they overpay for the house today. Discuss whether they should argue about paying for the home inspection and risk losing the house. Submit the best offer possible to get the best home your buyers want while still protecting their interest.

Keep the Process Smooth After Acceptance

As we all know, so much happens between offer acceptance and the Close of Escrow (COE). For each real estate transaction, everyone involved needs to understand what they expect to do, especially your clients. Explain what information buyers need to remove each contingency and commit to a firm intent to close escrow. For example, in California, the contingency removal date is typically 17 days from acceptance. In other states, such as Colorado, Florida, Georgia, Illinois, and Texas, the passage of the period itself removes the contingency. Also, at what point your clients would have to pay for fees they've agreed to cover. For instance, do buyers in your state have to pay the Homeowners Association (HOA) fees or the cost of the home inspection before closing? Do buyers pay for the appraisal upon ordering, or through the COE? When and how to send the wire, so they don't become victims of Wire Transfer Fraud? Let buyers know what they must pay, and when each bill is due. This way, they can allocate the money and have it available for each resource.

Educate your customers on Wire Transfer Fraud. Consider adding a Wire Transfer Disclosure to your email signature. Tell your clients not to click on any links in their emails. Hand-deliver the Wiring Instruction Form to your buyers and instruct them to send the wire to what you provided them and nowhere else. Otherwise, the safest way to do it is via a "cashier's check." Make sure they take no actions and to inform you anytime they receive communications regarding wiring instructions.

To prevent Impersonation Fraud and any attempts of an impostor requesting a fraudulent payment, instruct your buyers to add each vendor included in the transaction as a new contact on their phones. To avoid fraudsters, provide them with a list of contact information of all vendors, such as title agents, attorneys, escrow officers, and mortgage consultants. Introduce buyers to all people involved in the transaction early in the process and before them signing their initial disclosures.

When inspection reports come in, don't just call buyers and ask if they had questions. Attempt to sit down with them and go over any red flags. Also, send homebuyers reminders of each deadline and expected responsibilities, such as signing a contingency form, or an addendum to the sales agreement.

During the purchase period, your buyers must keep up with their obligations and make all their payments on time. Also, make clear that they must avoid any major purchases that might alter their credit score or cash on hand. Any adverse effect on their credit record or liquid assets might shake the closing of the deal. There is nothing more disappointing than getting to the closing table only to find out that the loan must go back to the underwriting department, because Bob had just purchased a brand-new vehicle, and Barbara just ordered a new bedroom set. Make sure you warn your buyers about the downfall of new purchases during the home buying transaction, just like you did inform them about Wire Transfer Frauds.

Moreover, if your buyers are getting gift money from a close relative for the down payment, George, the grantor, must source the gift funds. He may also need to file tax forms with the Internal Revenue Service (IRS) for gift amounts over 15,000 dollars. Ask your buyers, Bob and Barbara, to check with their accountant about State Tax Laws regarding Gift Funds.

Furthermore, call George, the relative, and warn him against Gift Money Fraud. The lender or investor may call him to verify that the given money was, in fact, a gift: *"Hello George, how will Bob pay you back?"* If he answers anything other than the money being a gift, then the entire transaction may fall apart.

When the process gets near closing, you need to manage your buyers' expectations of time frames. Explain the difference between funding and recording. How does funding on a Friday affect their payments? In California, signing documents doesn't mean you take occupancy right away. Recording the property can take a few working days—more if you run into official holidays. Let them know when to schedule for utilities to turn on, and the accounts transferred into their names.

Send Bob and Barbara a list of contact information of the utility companies. Make sure you have the information they will need handy from water, gas, trash, to electric, cable, and internet. Write a personal "Get Ready to Move" note. *"Hello, your house is closing in five business days. Make sure you call and transfer the utilities into your name the day your house closes — [insert the anticipated closing date]. For your convenience, we have included the contact information of all utility companies. Also, a list of vendors we trust and love. Leverage these third-party partners to ease the transition of moving to your new home, and so on!"* Give this letter and the list of vendors to both sellers and buyers so they can close out and transfer utility accounts with no hassle. Also, help them get ready to move to their new homes conveniently. Do whatever else you can to smooth the process and make their real estate experience an enjoyable, and pleasant journey. After all, **you are an incredibly useful and fun real estate professional people seek when they need to buy or sell homes.**

Note

All companies, including your real estate firm, should purchase a Cyber Insurance policy that covers any aspect of Cyber Crimes such as Hacking, Wire Transfer Frauds, or any other digital embezzlement. A General Liability or Errors and Omission (EO) doesn't cover neither protects you against digital crimes. An annual $400 investment into a Cyber Insurance policy can protect you and your clients against huge losses and misfortunes.

Take the Next Step

Execute Tactic 12.6: Create a Customized Homebuyer's Guide

What Happens Next?

"Growing your business isn't enough; you must dominate!"
—James Tyler

Domination

There are only three ways to grow any business: increase the number of customers, profit per transaction, and the frequency of which customers buy. Since you can't control how often customers sell or buy, you can lean on increasing your commission for more revenue. However, as a real estate professional, your commission relies solely on the number of customers you close. Closing zero customers generate zero dollars. Hence, **the most significant factor in growing your business is increasing the number of customers!**

You can increase the number of clients in several ways:

1. **Grow Your Database**: Escalate the momentum of your business by adding more people to your database.
2. **Increase Your Referral Sources**: Partner with more local businesses.
3. **Expand Your Team**: Recruit and hire top agents.

It will be difficult for you to hire more agents or partner with other companies if you are a small firm or new to the business. Without a robust business model and a stable stream of consistent leads, you'll have a hard time hiring top producing agents. Also, it wouldn't make sense for third-party vendors to partner with you if you have a small customer base as they rely on referrals too. Thus, you must focus on growing your database to leverage the latter two options—increasing your referral partners and expanding your sales force. Otherwise, if you're a top producer, leverage all three methods to gain more clients.

In this book, I explained how you could increase your mass by adding 6,000 contacts to your database all at once and therefore increase the momentum of your business quickly and exponentially. See chapter 7: BUILD MASSIVE MOMENTUM. Also, in chapter 9: GENERATE HIGH-QUALITY LEADS, I showed you how to get involved with your local community and become a celebrity real estate agent. With the rest of this book, I armed you with all the tools needed to grow your business and lead as one of the top real estate professionals in the nation. Next, let's cover domination, and how you can retain your position as a top producing real estate professional.

Innovate, Grow, and Dominate

Three factors that will take your business from being successful to being one that will innovate, grow, and dominate.

Value First

As of 2019, Tony Robbins' net worth is roughly 500 million dollars, Gary Vaynerchuk's is approximately $50 million, and Russell Brunson is about $37 Million. Many more entrepreneurs and business leaders are making millions of dollars. The one invaluable element that they all have in common is providing "Value First." Big companies like Facebook, Instagram, Amazon, Twitter, Uber, and many others made little money when they started. However, they made vast amounts of profit years down the road. Why? Because they provide value first, once they hook people, they monetize.

Vaynerchuk wrote an entire book about this subject in his "Jab Jab Jab Right Hook." To dominate your market place, you must provide value first. The age of "me, me, me" is over! It is all about the customer, not about you! What have you written, said, recorded, published, or done that is 100% about your customer, and that you are providing for free, or for pennies?

What expertise and benefits do you offer that can add instant value to your potential clients? Package those values in various formats to present to your audience. People consume information differently! Your diverse methods and quick delivery of your benefits will do wonders. For example, you can host in-person events and schedule one-on-one meetings. Also, write articles and record podcasts, videos, and webinars. You can leverage social media to create educational real estate buying and selling groups on Facebook or LinkedIn. Combine social media groups with your blog, podcast, and YouTube channel, and you will become an authoritative leader in your real estate space. Your community will eventually build you, your brand, and your business. See chapters 3, 4, and 5 to learn how you can convert your core values into valuable content to influence and dominate.

Move Early

The next strategic opening leverages an increasing pace of disruption. New battlegrounds in business are emerging at a sped up-tempo. If you can identify when and how your market will evolve, you can move to the next battlefield ahead of your opponents. Then, establish a defensive position before your competition even realizes that the future has changed.

Early in the game, Google realized that internet users would increasingly start their surfing sessions through a search engine. They realized that instead of typing an address into their browser's URL box, users would prefer to enter search terms into their favorite search engine. Google moved onto this battlefield early in the game and positioned them to play defense instead of offense. Despite costly efforts to improve their search engines, both MSN and Yahoo! continue losing ground to Google. It has turned the game on its main competitors.

Also, Walmart's success sprung from a simple initial tactic: identifying the next battleground, setting up a stronghold there, and waiting for the competition. While large retailers such as Sears and JCPenney positioned stores only in large cities and town centers, Walmart took the opposite approach: it focused on smaller towns. Walmart is thriving! Sears and JCPenney are out!

What does this mean to you as a real estate professional? Find areas with a turnover rate of at least 7%, source their data, prepare for growth, create awareness, market to them, farm offline and online, prospect, close, and dominate. See chapters 7 through 12 to learn how you can leverage data and technology to dominate any market.

Lock up Resources

Restrict your competitors' access to resources, preempting their ability to resist your expansion.

Before the iPod, hard drives were too large to fit in an appealing box. Toshiba had recently developed a revolutionary new hard drive the same size as a flash-memory. Each hard drive could hold ten times the number of songs flash-memory-based players can hold. When Apple launched its first iPod, it signed an exclusive agreement with Toshiba, which prevented competitors from following quickly. iPod's success was because of Apple purchasing Toshiba's entire inventory of these new hard drives to prevent competitors like Sony from following too closely. By locking up Toshiba's supply, at least temporarily, Apple made it impossible for competitors to match the iPod's performance. The exclusive agreement gave Apple a period of protection for several months, which can make a world of difference.

Many other companies have sustained abundant profits by applying this move. Coca-Cola, for instance, signed large, long-term supply contracts with corn syrup manufacturers. It blocked Pepsi from their supply of corn syrup. Also, oil and gas companies compete primarily by locking up drilling rights.

The merit of these case studies is that you, as a real estate professional, can also lock up resources. You can restrict other agents in your territory from accessing means whereby you make them only available to you. Use these resources to your advantage to dominate, and also to prevent other real estate agents from surpassing your growing benchmark. For instance, sign up exclusive partnership agreements with every local vendor, company, and storefront that comes to mind. Help send them business exclusively! In return, they would refer deals back to you! This tactic alone can suffice in generating you more business than any real estate agent in your area.

Lock up resources by reaching out to top influencers. I am not referring to a one-time Instagram shout out from an influencer with a few hundred thousand followers. I am talking about reaching hundreds of top influencers to introduce yourself, make a connection, find out ways to serve them, and eventually get into the inner circle of a few. Once you're in, you'll have tremendous opportunities to gain a reputable brand with national recognition. See chapter 9: GENERATE HIGH-QUALITY LEADS to learn how you can leverage multiple lead generation strategies to thrive and dominate.

Build Your Domination System

Here are the five steps to building a reliable and potent real estate domination system:

1. **Perfect your domination strategy**: Follow all the steps in this book to figure out which plans work best for you and your business. Execute the tactics and strategies with which you become most familiar. Test, measure, improve, and make them yours. Perfecting growth tactics shared in this book will help you build your unique domination system. They will become the engine that will drive your real estate business forward, regularly and consistently.

2. **Narrow your niche**: Craft one solution to the biggest, the most urgent, and the most critical problem your customers need to solve now. What can you tell me that makes me want to call you and hire you right now? Did you write your value proposition? For example, if you're going to be the agent that every seller in a particular city must work with, you need to brand yourself as The Home Sellers Expert. The biggest mistake most agents do is they try to be too general with the fear of losing out. Being a "Jack of all trades, master of none" leads to losing out on a ton of business! Narrow your niche and specialize so you can form a tribe. When people ask you if you work with buyers, sellers, FSBOs, or Expired Listings, pick only one. When they ask which territory you serve, you shouldn't say you serve the entire state or county. Be specific, down to a few cities at max.

3. **Master every area of your growth and domination system**: To dominate, you must be able to train and manage every new hire once you scale your business. You can't possibly teach if you haven't gone through the process yourself. If you don't understand business development, marketing, and sales, don't expect to grow your business, period! If you were to give a speech at a TEDx event, can you intelligently speak about these three departments and the five steps to domination? Do you understand the importance of demand generation, lead generation, activation, prospecting, retention, conversion, sales, and how each area can affect your real estate business? If the answer is no, go back and read the book several times until you're competent to be the captain of your ship.

4. **Gradually remove yourself from the system so you can scale and grow**: Once you master every aspect of the three most influential departments, remove yourself from each function. Hire a professional recruiting company to hire top real estate producers. Otherwise, bring in interns to train and eventually replace your position within those departments.

First, you need a closer! The closer can be an additional agent to handle the closing of the sales. Second, you need a marketer to manage appointment-scheduling for the closer. Also, you will need a business development representative to handle brand awareness, prospecting, and generate leads for the marketer who sets the appointments for the closer. Then, keep hiring as income grows until you automate every area of your domination system and your business.

5. **Finally, remove yourself altogether from the system so you can dominate your marketplace**: Review, manage, test, maintain, improve, and optimize the structure of your business to scale growth and dominate. Rinse and repeat!

The 10 P's to Domination

In a nutshell, "Go big, go all in, and take action now!" to amplify your performance, become a top producing real estate professional, and dominate your space. Follow these ten profound objectives in your overall business development, marketing, and sales strategies to achieve astounding business results:

1. *Plan*: *Plan Your Purpose and Priorities*

Envision the big picture! How does success look like to you? Continuously set and revisit your SMART[2] goals, if not daily, at least weekly. Keep your eyes on the prize! Plot out a strategic plan with specific milestones and objectives. Know your Key Performance Indicators (KPIs) and execute them. Start every day with your high leverage activities and the most important priorities to keep your business moving forward and upward.

2. *Push*: *Push Your Potential with Positivity and Passion*

Follow the 3 E's model for taking the initiative (Energy, Enthusiasm, and Endurance). Stay fired up, ambitious, and enthusiastic! Generate the power and endurance that will push you through every day. Create high levels of energy by staying positive, working out, hydrating, and eating healthy. Your positive energy not only will attract new clients but also positively influence others around you. Also, your positivity, passion, and massive potential have the power of improving your cognitive abilities, taking in information, retaining it, and acting on it. Thus, leverage enthusiasm and optimism as they are contagious and will eventually lead to exponential success.

3. *Position: Perception of Preeminence*

Position yourself and your business to dominate! It is essential to set high standards and maintain them. Whether they are the Standard Operating Procedures (SOPs) for your real estate office or your High Leverage Activities (HLAs) for yourself—determine those rules and abide by them. For example, adhere to a consistent number of prospecting calls each day, or the number of appointments set each week. Develop an office-wide minimum standard for active escrows or pending listings each month. More importantly, a standard for the minimum reserve amount inside your business checking account (e.g., 3-6 months minimum burn rate). Position your business to be reliable, operational, and financially healthy.

4. *Privacy: Personal and Private*

Create a personal time slot where no one distracts you. Block out an hour each day to call prospects or follow up on important matters. Nevertheless, maintain quiet time where you eliminate distractions and focus on the most critical high leverage dollar-producing activities.

5. *Process: Points of Process*

Establish a clear sales management strategy to support your customers and the sales team. Develop a systematic process for your clients and your team that clarifies each point of the buying and selling milestones to close more transactions. Ensure that everyone is in sync through each level of the sales process. Communicate the next steps to your customers. Create a series of short videos for every stage of the selling and buying journey to drip to your clients through emails and SMS. Streamlining the process eases customer pain and reduces frustration and stress levels. Educate your clients on the process so you can better prepare them, instead of them having to guess or anticipate what comes next. Always be refining, optimizing, and improving to accelerate sales and growth. Enhancing every point of the real estate process is a critical part of your domination system. Invest the time and energy into perfecting this piece.

6. *Progress: Perceive Your Progress*

Know your numbers and measure everything that matters! Measure calls, emails, text messages, contacts, leads, open houses, appointments, and monthly sales. Remember, if you can't measure it, you can't improve it. Develop specific sales activity metrics for yourself and each salesperson in your office. Follow both the Qualitative and Quantitative approaches to know the numbers and pivot according to the data at hand. Don't act and run your business solely based on your gut feeling or charisma, but on hard data. Create a repeatable and scalable business by knowing your numbers and the math behind success. Use data to make better and informed decisions towards your real estate future.

7. *Practice*: *Practice Your Playbook*

Practice and roleplay your scripts daily! The more you practice various scenarios, the better you'll get, and the more natural you'll sound. The more confident you are, the more clients will trust you, and the more sales you'll close. Top real estate producers practice their craft daily, and so should you.

8. *Prospect*: *Propel Your Pipeline*

Increase business momentum and the number of closed sales by creating a prospecting consistency. Improve pipeline prospecting, planning, and reporting by using various methods of prospecting such as text messaging, emails, phone calls, voicemail drops, and paid ads. Pick the prospecting methods that work for you and stick to them.

9. *Present*: *Portray the Picture*

Persistently set more appointments, and continuously get better at negotiation. It would be best if you were out on interviews and meetings with clients weekly. If you are a new agent, you must practice your presentation skills. It doesn't have to be to clients only; you can ask to do so with a colleague, a family member, or a friend. Master the art of presenting by making it a weekly habit. Continuously get feedback and improve your presentation skills.

10. *Persuade*: *Push the Limits*

Follow up, train, and recruit! Follow up with prospects, clients, and people within your network using the 3 F's model: follow-up, follow-through, and follow-back. Train and transfer your skills to other people on your team so you can scale, grow, and dominate. Recruit new talent by implementing a sales management approach that lays the foundation for larger sales teams. Roll out a hiring process that attracts only top players. Then, persuade them to become the success you've created.

#WhatsNext

Expanding your real estate business is an excellent path to growth. However, to dominate, you must create a big name for yourself! You and your company must create a significant impact on the market, acquiring quality leads in mass, and generating massive momentums much faster.

> *Take the Next Step*
>
> *Execute Tactic 13.0: Research the Top 250 National Real Estate Professionals*

This book has provided you with all the secrets, tools, methods, tactics, and strategies you need to become a top-producing real estate agent and dominate your marketplace. All you need to do is believe in yourself and execute. Get help! You don't need to do everything yourself.

To become the number one producer and dominate, you need to know where you stand. Once you look into at least the top 100 dominating agents, consider answering the (BCC-MMV) questions:

1. **Bench-Marketing:** How does your brokerage compare to others in your region and nationally?
2. **Coaching:** Are you leveraging mentoring and coaching services to recruit? Are you training new talent? Are you improving your operations and Structured Operating Procedures (SOPs)?
3. **Commission Concepts:** Have you ever wondered what your company dollar would look like with different commission plans in place?
4. **Market Share Analysis:** How can you take advantage of market data to provide you with a strategic advantage?
5. **Mergers and Acquisitions:** Have you considered selling your business or acquiring another firm to expand, grow, and dominate?
6. **Valuations:** What is your brokerage worth in today's market?

"If you can dream it, you can do it."
—Walt Disney

#GetInvolved & #StayInvolved

#GetInvolved: In this book, I have mentioned that to succeed, you need to become a thoughtful legacy leader that provides value in every aspect of the business, daily, offline, and online!

#StayInvolved: I have taught you how to build a sustainable, profitable system with every moving part of the three crucial departments: Business development, marketing, and sales. Keep implementing and improving those strategies daily.

Growing your business isn't enough; you must dominate! Make sure you have executed all the strategies and tactics outlined in this book. Rinse and Repeat!

"Tell me, and I forget, teach me, and I remember, involve me, and I learn."
—Benjamin Franklin

Take the Next Step
Execute Tactic 13.1: Build Your Domination System

#UncoverHiddenPearls

From a successful growth mindset to better management, from business development to marketing, sales, and growth; no one can stop you from succeeding and dominating your real estate market.

Customers are the main drivers of any business. Think of potential clients as pearls concealed on the deep ocean floor. Floating on the surface and pursuing real estate as a part-time gig will not get you precious gems. You must commit and give it your 100%. You have to provide it with everything in your power to drive that momentum. You have to think big and dig deep into the profound sea. Explore to unveil every hidden shell and pearl that will make your business shine under the sunlight of success and domination.

It is time to stop reading this book and time to create a big name for YOU.

I wish you the best of practice, and thank you for following along!

Please continue your domination journey at *JamesTyler.com*.

"Domination is a process; practice it to master it!"
—James Tyler

REFERENCES

PREFACE

- "2017 Real Estate Forecast Breakfast." *Bend Chamber of Commerce*, 20 Apr. 2017, bendchamber.org/bend-event/2017-real-estate-forecast-breakfast/.
- Carlsen, Magnus. "Magnus Carlsen." *Wikipedia*, Wikimedia Foundation, 16 June 2019, en.wikipedia.org/wiki/Magnus_Carlsen.
- "Chess." *Wikipedia*, Wikimedia Foundation, 25 June 2019, en.wikipedia.org/wiki/Chess.
- INTRODUCTION
- Carlock, Byron. "Emerging Trends in Real Estate® - US and Canada 2019." *PwC*, 2019, pwc.com/us/en/industries/asset-wealth-management/real-estate/emerging-trends-in-real-estate.html.
- Carlock, Byron. "Real Estate 2020 Building the Future ." PwC Real Estate 2020: Building the Future, *PwC*, 2014, pwc.com/sg/en/real-estate/assets/pwc-real-estate-2020-building-the-future.pdf.
- "Wikimedia List Article." *Wikiquote*, Wikimedia Foundation, Inc., 23 May 2019, en.wikiquote.org/wiki/Chinese_proverbs.

CHAPTER 1: BECOME A LEGACY LEADER

- Assadi, Arman. "A 9-Step Framework for Creating a Morning Ritual." *Entrepreneur,* 14 Jan. 2015, entrepreneur.com/article/241691.
- Douwes, Alexandra. "5 Tips For Creating A Morning Ritual That Sticks." *Forbes,* Forbes Magazine, 28 June 2018, forbes.com/sites/alexandradouwes/2018/06/25/5-tips-for-creating-a-morning-ritual-that-sticks/#596440cc5e39.
- "Keller Center Research Report." Keller Center Research Report, *Baylor University,* Mar. 2018, baylor.edu/business/kellercenter/doc.php/306999.pdf.
- "Outliers Quotes by Malcolm Gladwell." *Goodreads,* Goodreads, 2019, goodreads.com/work/quotes/3364437-outliers-the-story-of-success.
- "Wayne Gretzky Quotes." *BrainyQuote,* Xplore, 21 June 2019, brainyquote.com/quotes/wayne_gretzky_378694.
- "Winston Churchill Quotes." *BrainyQuote,* Xplore, brainyquote.com/quotes/winston_churchill_103739.

CHAPTER 2: ARTICULATE DESIRES INTO TANGIBLE GOALS

- "Applying Quantitative and Qualitative Measurement to Your Social Media." Applying Quantitative and Qualitative Measurement to Your Social Media Strategies, *Hootsuite,* 2019, hootsuite.com/education/courses/social-marketing/strategy/measurement.
- Covey, Stephen R. "The 7 Habits of Highly Effective People." *FranklinCovey,* Free Press, 9 Nov. 2004.
- Jeary, Tony. "High Leverage Activities (HLAs)." Tony Jeary International - The Results Guy, tonyjeary.com/wp-content/uploads/2018/06/TonyJearyHLAWhitepaper.pdf.
- "Lewis Carroll Quotes." *BrainyQuote,* Xplore, brainyquote.com/quotes/lewis_carroll_165865.
- "Pareto Principle." *Wikipedia,* Wikimedia Foundation, 10 June 2019, en.wikipedia.org/wiki/Pareto_principle.
- "Parkinson's Law." *Wikipedia,* Wikimedia Foundation, 30 Apr. 2019, en.wikipedia.org/wiki/Parkinson%27s_law.
- Peter, Drucker F. The Effective Executive: the Definitive Guide to Getting the Right Things Done. HarperCollins US, 2011.
- S, Surbhi. "Difference Between Qualitative and Quantitative

Data (with Comparison Chart)." *Key Differences,* 3 Nov. 2017, keydifferences.com/difference-between-qualitative-and-quantitative-data.html.

- "SMART Criteria." *Wikipedia,* Wikimedia Foundation, 9 June 2019, en.wikipedia.org/wiki/SMART_criteria.
- Suter, W. Newton. "Qualitative Data, Analysis, and Design." A Critical Thinking Approach, Second Edition, Sage Publications, 2019, sagepub.com/sites/default/files/upm-binaries/43144_12.pdf.

CHAPTER 3: DISCOVER YOUR CORE VALUES

- Belludi, Nagesh. "Albert Mehrabian's 7-38-55 Rule of Personal Communication." Right Attitudes, 27 Oct. 2017, rightattitudes.com/2008/10/04/7-38-55-rule-personal-communication.
- "Credibility." *Wikipedia,* Wikimedia Foundation, 11 June 2019, en.wikipedia.org/wiki/Credibility.
- Dib, Allan. "The Riches Are In the Niches." *Successwise,* 30 June 2018, successwise.com/the-riches-are-in-the-niches.
- Galden, Dane L. Mortgage Lending Principles & Practices. *Hondros Learning.,* 2007.
- Hajli, M. Nick. "A Study of the Impact of Social Media on Consumers." *International Journal of Market Research,* vol. 56, no. 3, 2014, pp. 387–404., doi:10.2501/ijmr-2014-025.
- Narus, James C. AndersonJames A. "Business Marketing: Understand What Customers Value." *Harvard Business Review,* 1 Aug. 2014, hbr.org/1998/11/business-marketing-understand-what-customers-value.
- "Og Mandino Quotes." *BrainyQuote,* Xplore, 21 June 2019, brainyquote.com/quotes/og_mandino_885199.

CHAPTER 4: ELEVATE THE POWER OF YOU

- "A Quote by Bob Marley." *Goodreads,* Goodreads, 2019, goodreads.com/quotes/884474-you-never-know-how-strong-you-are-until-being-strong.
- "A Quote by F. Burton Howard." *Goodreads,* Goodreads, 2019, goodreads.com/quotes/343684-if-you-want-something-to-last-forever-you-treat-it.
- Buechner, Frederick. *Telling Secrets.* HarperSanFrancisco, 1991.
- Da Cunha, Margot. "6 Storytelling Tips to Tell Your Business Story Like a TED Pro." *WordStream,* 8 Nov. 2018, wordstream.

com/blog/ws/2014/12/05/business-storytelling.

- Deverell, Eva. "Creative Writing Blog." *E.a. Deverell,* 2019, eadeverell.com/.
- "MoSCoW Method." *Wikipedia,* Wikimedia Foundation, 28 Apr. 2019, en.wikipedia.org/wiki/MoSCoW_method.
- Lim, Brian. "Social Marketing for Real Estate Agents." Z57, 3 Jan. 2019, blog.z57.com/page/8.
- Sapolsky, Robert M. "'Us vs. Them' Thinking Is Hardwired-But There's Hope for Us Yet." *Big Think,* Big Think, 8 Feb. 2019, bigthink.com/videos/robert-sapolsky-us-vs-them-thinking-is-hardwired-but-theres-hope-for-us-yet.
- "Unique Selling Proposition." *Wikipedia,* Wikimedia Foundation, 10 June 2019, en.wikipedia.org/wiki/Unique_selling_proposition.
- Wrangham, Richard. "Insights Into the Brain, in a Book You'Ll Wish You Had in College." *The New York Times,* 6 July 2017, nytimes.com/2017/07/06/books/review/behave-robert-m-sapolsky-.html.

CHAPTER 5: SPEAK YOUR MESSAGE WITH INFLUENTIAL CONTENT

- "Best 43 Call To Action Phrases Proven to Insanely Increase Clicks and Leads." *Hello Bar,* 21 June 2018, hellobar.com/blog/call-to-action-phrases.
- Gyant, Nia. "Why Is Social Media Important For Your Business?" *ThriveHive,* 22 Feb. 2018, thrivehive.com/social-media-important-business.
- Jackson, Dominique. "Call to Action Phrases That Will Convert." *Sprout Social,* 17 May 2019, sproutsocial.com/insights/call-to-action-phrases/.
- Miller, Donald. *Building a Storybrand: Clarify Your Message so Customers Will Listen.* HarperCollins Leadership, 2017.
- Southern, Tom. "Why These 21 Headlines Went Viral (and How You Can Copy Their Success)." *Optinmonster,* 7 Jan. 2019, optinmonster.com/why-these-21-headlines-went-viral-and-how-you-can-copy-their-success.

CHAPTER 6: SHARPEN YOUR AXE

- "Audience Network." *Facebook Business,* 2019, facebook.com/business/marketing/audience-network.

- Editor, Business News Daily. "15 LinkedIn Marketing Hacks to Grow Your Business." *Business News Daily,* 29 Sept. 2014, businessnewsdaily.com/7206-linkedin-marketing-business. html.
- "Email Marketing, Email Design & Email Testing Tools." *Litmus,* 2019, litmus.com/.
- "Facebook Messenger for Business." *Facebook Business,* 2019, facebook.com/business/marketing/messenger.
- Garson. "To Cut Down a Tree in Five Minutes Spend Three Minutes Sharpening Your Axe." *Quote Investigator,* 29 Mar. 2014, quoteinvestigator.com/2014/03/29/sharp-axe/.
- *Google Search Console,* Google, search.google.com/search-console/about.
- "Introducing the Online Real Estate Software System." *MoreSolds Contact Manager for Real Estate Agents,* 2019, moresolds. com/Features_Scheduling-Tools.asp.
- "Social Media Management Software for Growing Brands." *Buffer,* 2019, buffer.com/.
- "Text Messaging for Real Estate." *SMS Marketing & Text Marketing Services* – Try It For Free, 2019, simpletexting.com/ guide/text-messaging-for-real-estate/.
- Tokareva, Julia. "How Virtual Reality Is Changing The World Of Advertising." *Forbes,* Forbes Magazine, 6 Dec. 2017, forbes. com/sites/quora/2017/12/06/how-virtual-reality-is-changing-the-world-of-advertising/#f0182a168d51.
- Vijayakumar, Adarsh. "Augmented Reality." *YouTube, YouTube,* 25 Oct. 2018, youtube.com/watch?v=I38wBVT52oU.

CHAPTER 7: BUILD MASSIVE MOMENTUM

- "Best Practices Report | Improving Data Preparation for Business Analytics." *Transforming Data with Intelligence,* 1 July 2016, tdwi.org/research/2016/07/best-practices-report-improving-data-preparation-for-business-analytics.aspx.
- Clarke, Todd. "22+ Instagram Statistics That Matter to Marketers in 2019." *Hootsuite Social Media Management,* 5 Mar. 2019, blog. hootsuite.com/instagram-statistics/.
- *Dennis Gabor > Quotes > Quotable Quote, Goodreads,* 2019, goodreads.com/quotes/7733339-the-future-cannot-be-predicted-but-futures-can-be-invented.
- "Farming for the Future." *Chime Technologies,* 19 Oct. 2017,

chime.me/blog/farming-for-the-future.

- "Global Social Media Ranking 2019 | Statistic." *Statista*, 2019, statista.com/statistics/272014/global-social-networks-ranked-by-number-of-users/.
- Hinton, Todd. "What Is Data Enrichment?" *RedPoint Global*, 24 July 2018, redpointglobal.com/blog/what-is-data-enrichment/.
- "How Are Velocity and Speed Related." *Answers*, Answers Corporation, 2019, answers.com/Q/How_are_velocity_and_speed_related.
- "Instagram, Snapchat Duel for Millennial Attention." *EMarketer*, 10 Aug. 2017, emarketer.com/Article/Instagram-Snapchat-Duel-Millennial-Attention/1016316.
- "Momentum." *Wikipedia*, Wikimedia Foundation, 18 June 2019, en.wikipedia.org/wiki/Momentum.
- "Q&A For Telemarketers & Sellers About DNC Provisions in TSR." *Federal Trade Commission*, 30 Apr. 2019, ftc.gov/tips-advice/business-center/guidance/qa-telemarketers-sellers-about-dnc-provisions-tsr#payingforaccess.
- "Using Every Door Direct Mail." *USPS*, 2019, usps.com/business/every-door-direct-mail.htm.
- "Validate American Format of Phone Number." *Stack Overflow*, stackoverflow.com/questions/5841888/validate-american-format-of-phone-number.
- Yera, Adam. "Geographic Farming: Top Producer's Guide to Neighborhood Domination." *CompleteAgent*, 19 Apr. 2018, completeagent.io/realtors-guide-to-geographic-farming/.

CHAPTER 8: PLANT MORE SEEDS

- "A Quote by Benjamin Franklin." *Goodreads*, Goodreads, 2019, goodreads.com/quotes/103000-without-continual-growth-and-progress-such-words-as-improvement-achievement.
- "Best Practices Report | Improving Data Preparation for Business Analytics." *Transforming Data with Intelligence*, 1 July 2016, tdwi.org/research/2016/07/best-practices-report-improving-data-preparation-for-business-analytics.aspx.
- Blair, Ian. "Powerful Real Estate Marketing Ideas From 19 Top Experts." *BuildFire*, 12 Aug. 2018, buildfire.com/real-estate-marketing-ideas.
- Bourgeois, Diana. "Real Estate Marketing & Sales Strategy From the Pros." *Fit Small Business*, 2019, fitsmallbusiness.com/

real-estate.

- Burrowes, Brianne. "How Planting a Seed Can Change Your Life." *Tiny Buddha*, 22 Mar. 2014, tinybuddha.com/blog/plant-a-seed-change-your-life.
- Clarke, Todd. "22+ Instagram Statistics That Matter to Marketers in 2019." *Hootsuite Social Media Management*, 5 Mar. 2019, blog. hootsuite.com/instagram-statistics/.
- "Demand Generation." *Wikipedia*, Wikimedia Foundation, 2 June 2019, en.wikipedia.org/wiki/Demand_generation.
- *Dennis Gabor > Quotes > Quotable Quote, Goodreads*, 2019, goodreads.com/quotes/7733339-the-future-cannot-be-predicted-but-futures-can-be-invented.
- Elmo Lewis, Elias. "AIDA Sales Funnel." *ProvenModels*, 2019, provenmodels.com/547/aida-sales-funnel/.
- Farfan, Barbara. "Q&A For Telemarketers & Sellers About DNC Provisions in TSR." *Federal Trade Commission*, 30 Apr. 2019, ftc. gov/tips-advice/business-center/guidance/qa-telemarketers-sellers-about-dnc-provisions-tsr#payingforaccess.
- "Farming for the Future." *Chime Technologies*, 19 Oct. 2017, chime.me/blog/farming-for-the-future.
- "Global Social Media Ranking 2019 | Statistic." *Statista*, 2019, statista.com/statistics/272014/global-social-networks-ranked-by-number-of-users.
- Hinton, Todd. "What Is Data Enrichment?" *RedPoint Global*, 24 July 2018, redpointglobal.com/blog/what-is-data-enrichment/.
- "How Are Velocity and Speed Related." *Answers*, Answers Corporation, 2019, answers.com/Q/How_are_velocity_and_speed_related.
- "Instagram, Snapchat Duel for Millennial Attention." *EMarketer*, 10 Aug. 2017, emarketer.com/Article/Instagram-Snapchat-Duel-Millennial-Attention/1016316.
- "Lao Tzu Quotes." *BrainyQuote*, Xplore, 2019, brainyquote.com/quotes/lao_tzu_118184.
- "Managing Your Account | Facebook Help Center | Facebook." Facebook, 2019, facebook.com/help/messenger-app/165294433944588/1033919196669390.
- "Momentum." *Wikipedia*, Wikimedia Foundation, 18 June 2019, en.wikipedia.org/wiki/Momentum.
- "Q&A For Telemarketers & Sellers About DNC Provisions in TSR." *Federal Trade Commission*, 30 Apr. 2019, ftc.gov/tips-advice/business-center/guidance/qa-telemarketers-sellers-about-dnc-

provisions-tsr#payingforaccess.

- "Quick Real Estate Statistics." *nar.realtor,* 11 May 2018, nar. realtor/research-and-statistics/quick-real-estate-statistics.
- "Real Estate in a Digital Age 2017 Report." *National Association of REALTORS Research Department,* 2017, nar.realtor/sites/default/ files/reports/2017/2017-real-estate-in-a-digital-age-03-10-2017. pdf.
- "Realtors Property Resource." *Realtors Property Resource,* 2019, narrpr.com/.
- "Technology Survey." *National Association of REALTORS Research Group,* Sept. 2018, nar.realtor/sites/default/files/ documents/2018-technology-survey-09-18-2018.pdf.
- "Using Every Door Direct Mail." *USPS,* 2019, usps.com/ business/every-door-direct-mail.htm.
- "Validate American Format of Phone Number." *Stack Overflow,* stackoverflow.com/questions/5841888/validate-american-format-of-phone-number.
- Yera, Adam. "Geographic Farming: Top Producer's Guide to Neighborhood Domination." *CompleteAgent,* 19 Apr. 2018, completeagent.io/realtors-guide-to-geographic-farming/.

CHAPTER 9: GENERATE HIGH-QUALITY LEADS

- "Android v IOS Market Share 2019." *DeviceAtlas,* 6 June 2019, deviceatlas.com/blog/android-v-ios-market-share.
- "Creating and Placing High-Impact Advertisements – Part 2: Writing High-Impact Advertising Copy." *Real Estate Training and Coaching Real Estate Champions,* 2019, realestatechampions. com/articles/real-estate-marketing/creating-and-placing-high-impact-advertisements-part-2-writing-high-impact-advertising-copy.
- "Global Cloud-Based Communications and PR Solutions Leader." *Cision,* 17 Feb. 2015, www.cision.com/us/2015/02/ integrated-promotion-how-to-build-a-strategy-that-increases-traffic-visibility/.
- Kuenn, Arnie. "Don't Settle for Organic; Amplify Your Content." *Content Marketing Institute,* 22 Aug. 2017, contentmarketinginstitute.com/2017/08/content-amplification-promote-distribute.
- "Law of Attraction (New Thought)." *Wikipedia,* Wikimedia

Foundation, 19 June 2019, en.wikipedia.org/wiki/Law_of_attraction_(New_Thought).

- Lenouvel, Jess. "Want to Hit 7-Figures in 2019?" *Go.thelistingslab.com*, 2019, go.thelistingslab.com/.
- "List of Cognitive Biases." *Wikipedia*, Wikimedia Foundation, 17 June 2019, en.wikipedia.org/wiki/List_of_cognitive_biases.
- Marrs, Megan. "22 Low-Budget Marketing Ideas For Small Businesses." *WordStream*, 9 June 2019, wordstream.com/blog/ws/2014/10/01/marketing-ideas-for-small-businesses.
- Newberry, Christina. "Social Media Advertising 101: How to Get the Most Out of Your Ad Budget." *Hootsuite Social Media Management*, 6 May 2019, blog.hootsuite.com/social-media-advertising.
- "Online Advertising." *Wikipedia*, Wikimedia Foundation, 20 June 2019, en.wikipedia.org/wiki/Online_advertising.
- Pradhan, Advait. "Beacon-Based Contextual Marketing Opportunities in the Real Estate Industry." *Softwebsolutions*, 31 Jan. 2019, softwebsolutions.com/resources/beacon-contextual-marketing-in-real-estate-industry.html.
- "Push Notifications Explained." *Airship*, 2019, airship.com/resources/explainer/push-notifications-explained/.
- Sara. "Blue Car Syndrome." *Wordy Evidence of the Fact*, 1 Jan. 1970, wordyevidenceofthefact.blogspot.com/2010/09/blue-car-syndrome.html.
- Singh, Swagat. "Push Notifications Are Enabling Real Business Results, Are They?" *Complex Challenges. Simple Solutions*, 2019, innovapptive.com/blog/push-notifications-enabling-real-business-results.
- "Snowball Effect." *Wikipedia*, Wikimedia Foundation, 16 Aug. 2018, en.wikipedia.org/wiki/Snowball_effect.
- "Urban Airship Expands to Web Push Notifications, Enabling Marketers and Developers to Send Personalized, Real-Time Messaging to Website Visitors." *MarketWatch*, 15 Mar. 2017, marketwatch.com/press-release/urban-airship-expands-to-web-push-notifications-enabling-marketers-and-developers-to-send-personalized-real-time-messaging-to-website-visitors-2017-03-15-9195052.
- "Using Every Door Direct Mail." *USPS*, 2019, www.usps.com/business/every-door-direct-mail.htm.
- "Vance Havner Quotes." *BrainyQuote*, Xplore, www.brainyquote.com/quotes/vance_havner_105323.

- "What Is a Baby Boomer?" *The Baby Boomers Club,* 21 Dec. 2017, babyboomersclub.org/what-is-a-baby-boomer.

CHAPTER 10: NURTURE MEANINGFUL RELATIONSHIPS

- "Ad Exchange." *Wikipedia,* Wikimedia Foundation, 16 Jan. 2019, en.wikipedia.org/wiki/Ad_exchange.
- Collins, Jim. "The Flywheel Effect." *Jim Collins - Articles - The Flywheel Effect,* 2019, www.jimcollins.com/article_topics/articles/the-flywheel-effect.html.
- "Create Your First Display Remarketing Campaign - Previous." *Google Ads Help,* Google, 2019, support.google.com/google-ads/answer/3210317?co=ADWORDS.IsAWNCustomer%3Dfalse&hl=en.
- "Customer Lifetime Value." *Wikipedia,* Wikimedia Foundation, 23 Apr. 2019, en.wikipedia.org/wiki/Customer_lifetime_value.
- "Data Leaders Weave An Insights-Driven Corporate Fabric." *Forrester,* 18 Dec. 2017, forrester.com/report/Data+Leaders+Weave+An+InsightsDriven+Corporate+Fabric/-/E-RES140571.
- "DoubleClick Market Share and Web Usage Statistics." *SimilarTech,* 2019, www.similartech.com/technologies/doubleclick.
- Ellis, Sean. "What Is a North Star Metric?" *Growth Hackers,* Growth Hackers, 5 June 2017, blog.growthhackers.com/what-is-a-north-star-metric-b31a8512923f.
- Galavan, Ruairi. "Activating Customers by Unlocking the Right Steps." *Inside Intercom,* 6 Mar. 2019, www.intercom.com/blog/activating-customers-long-term.
- "JM Storm Quotes." *Goodreads,* Goodreads, 2019, goodreads.com/author/quotes/18572951.JM_Storm.
- Muhammad, Fahad. "Retargeting 101: Everything You Need to Get Started and Achieve Greater ROI." *Instapage,* 16 July 2018, instapage.com/blog/what-is-retargeting.
- Rock, David, and Josh Davis. "4 Steps to Having More 'Aha' Moments." *Harvard Business Review,* 12 Oct. 2016, hbr.org/2016/10/4-steps-to-having-more-aha-moments.
- Seth, Shobhit. "13 Steps of a Real Estate Closing." *Investopedia,* Investopedia, 19 May 2019, investopedia.com/articles/mortgages-real-estate/10/closing-home-process.asp.
- "User Experience – What to Expect in 2019?" *Brandwill Agency,*

5 Feb. 2019, brandwillagency.com/user-experience-what-to-expect-in-2019.

- "UTM Parameters." *Wikipedia*, Wikimedia Foundation, 21 June 2019, en.wikipedia.org/wiki/UTM_parameters.
- Weintraub, Elizabeth. "Follow These Steps and Know What to Expect When You Sell Your Home." *The Balance*, The Balance, 10 Jan. 2019, thebalance.com/home-selling-path-a-to-z-1799042.
- "What Is ReTargeting and How Does It Work?" *Retargeter*, 2019, retargeter.com/what-is-retargeting-and-how-does-it-work/.
- Wu, Susan. "Activate or Die: 3 Keys to User Activation for SaaS (Part 1)." *500 Startups*, 6 May 2015, 500.co/activate-or-die-3-keys-to-user-activation-for-saas-part-1.

CHAPTER 11: CREATE PROMISING OPPORTUNITIES

- "Emotional Intelligence." *Wikipedia*, Wikimedia Foundation, 15 June 2019, en.wikipedia.org/wiki/Emotional_intelligence.
- Ferry, Tom. "5 Superpowers of Top Successful Real Estate Professionals | #TomFerryShow." *YouTube*, YouTube, 20 Mar. 2018, www.youtube.com/watch?v=55eXnJBhajk.
- "Formula For Success: Rise Early, Work Hard, Strike Oil (J. Paul Getty) Essay." *Bartleby*, 10 Oct. 2016, bartleby.com/essay/Formula-For-Success-Rise-Early-Work-Hard-FK964BXYLCXQ.
- Greene, John. "How to Get Your Prospects on the Yes Ladder (and Make the Sale)." *PhoneBurner Blog*, 19 Dec. 2017, www.phoneburner.com/blog/how-to-prime-prospects-to-say-yes.
- Heyl, Tim. "Session #15 Scripts: Introduction to Your Coach." *Kw MAPS Coaching*, 2015, media.realgeeks.com/uploads/tim_heyl_scripts.pdf.
- Hogan, Kevin. "Hypnotic Language Pattern Interrupt Reversal Technique w Kevin Hogan." *YouTube*, YouTube, 13 Sept. 2016, www.youtube.com/watch?v=jGcbxHh9njA.
- "Insanity Is Doing the Same Thing Over and Over Again and Expecting Different Results." *Quote Investigator*, 2019, quoteinvestigator.com/2017/03/23/same/.
- "J. Paul Getty Quotes." *BrainyQuote*, Xplore, 2019, www.brainyquote.com/quotes/j_paul_getty_100065.
- Kaufman, Aaron. "Keller Williams Realty's MAPS Coaching: Real Estate Agent Script of The Month - For Sale By Owners." *Real Estate Careers at Keller Williams Realty*, 19 June 2008, moving-

careers.com/keller-williams-realtys-maps-coaching-real-estate-agent-script-of-the-month-for-sale-by-owners.

- Roe, Helen. "Why a Pattern Interrupt Is Just What You Need." *HuffPost*, HuffPost, 6 Dec. 2017, www.huffpost.com/entry/why-a-pattern-interrupt-i_b_8075800.
- Serhant, Ryan. SELL IT LIKE SERHANT: *How to Sell More, Earn More, and Become the Ultimate Sales Machine*. HACHETTE BOOKS, 2019.
- Taylor, Madisyn. "Emotional vs. Intellectual Intelligence." *DailyOM*, 2019, www.dailyom.com/cgi-bin/display/articledisplay.cgi?aid=491.
- "'Conversational' Circle Prospecting Script." *CLME Certified Local Market Expert*, 2015, www.clme.com/wp-content/uploads/2013/08/Prospecting-Scripts.pdf.

CHAPTER 12: SEAL THE DEAL

- "2019 Profile of Home Staging." *National Association of REALTORS Research Group*, 14 Mar. 2019, www.nar.realtor/sites/default/files/documents/2019-profile-of-home-staging-03-14-2019.pdf.
- Bray, Ilona, et al. "Nolo's Essential Guide to Buying Your First Home." *Amazon*, NOLO, 28 Dec. 2012, www.amazon.com/Nolos-Essential-Guide-Buying-First/dp/1413317626.
- "Buyer Checklists." *AmeriTitle*, 2019, www.amerititle.com/Resources/WebDocs/AmeriTitle%20Buyer%20Checklist.pdf.
- Carlock, Byron. "Emerging Trends in Real Estate® - US and Canada 2019." *PwC*, 26 June 2019, www.pwc.com/us/en/industries/asset-wealth-management/real-estate/emerging-trends-in-real-estate.html.
- "Creating a Dynamic Listing Presentation." *Real Estate Training and Coaching | Real Estate Champions*, www.realestatechampions.com/articles/business-building/creating-a-dynamic-listing-presentation.
- Ferry, Tom. "Mindset, Model and Marketing!: The Proven Strategies to Transform and Grow Your Real Estate Business EBook: Tom Ferry: Kindle Store." *Amazon*, Amazon, 7 Aug. 2017, www.amazon.com/Mindset-Model-Marketing-Strategies-Transform-ebook/dp/B074N4NQFR.
- Hajli, M Nick. "A Study of the Impact of Social Media Consumers." *International Journal of Market Research*, 25 Jan. 2013, www.lyfemarketing.com/wp-content/uploads/2015/05/Compressed-PDF.pdf.

- Kerr-Dineen, Luke. "The 48 Greatest Quotes about Winning." *USA Today,* Gannett Satellite Information Network, 24 June 2019, ftw.usatoday.com/2016/02/best-sports-quotes-about-winning.
- "Leverage the Power of the Internet to Sell More Homes, Win More Listings and Attract More Buyers for Your Real Estate Property Listing." *ArchAgent,* archagent.com/Agent-Resources-Win-More-Listings.aspx.
- Mashore, Krista. "Ready to Be a Real Estate Powerhouse?" *Krista Mashore,* 2019, kristamashore.com/.
- Pawar, Amruta Vijay. "Study of the Effectiveness of Online Marketing on Integrated Marketing Communication." *DY Patil University,* Nov. 2014, www.dypatil.edu/schools/management/wp-content/uploads/2015/05/Study-Of-The-Effectiveness-Of-Online-Marketing-On-Integrated-Marketing-Communication-Amruta-Pawar.pdf.
- Pratt, Hoss. "LISTING BOSS: The Definitive Blueprint For Real Estate Success EBook: Hoss Pratt: Kindle Store." *Amazon,* Amazon, 29 Mar. 2017, www.amazon.com/LISTING-BOSS-Definitive-Blueprint-Success-ebook/dp/B06XXWWFRQ.
- "Real Estate Listing Presentation." *Placester,* 7 Mar. 2018, placester.com/resources/real-estate-listing-presentation-template.
- Seiler, Michael J., et al. "Toward an Understanding of Real Estate Homebuyer Internet Search Behavior: An Application of Ocular Tracking Technology." *SSRN,* 28 Mar. 2012, papers.ssrn.com/sol3/papers.cfm?abstract_id=2029823.
- Smaby, John. "Home Buyer and Seller Generational Trends." *National Association of REALTORS,* 3 Apr. 2019, www.nar.realtor/research-and-statistics/research-reports/home-buyer-and-seller-generational-trends.
- "Technology Survey." *National Association of REALTORS Research Group,* Sept. 2018, www.nar.realtor/sites/default/files/documents/2018-technology-survey-09-18-2018.pdf.
- "The All-in-One Social Media Strategy Workbook." *Hootsuite,* socialbusiness.hootsuite.com/rs/hootsuitemediainc/images/Social%20Media%20Strategy%20Workbook.pdf.
- Treece, Kiah. "Comparative Market Analysis: Ultimate Guide to a CMA in Real Estate." *Fit Small Business,* 15 Mar. 2019, fitsmallbusiness.com/comparative-market-analysis.
- "Trial Closes, Tie Downs, and Gaining Agreements." *National Association of Expert Advisors,* 2 Dec. 2016, naea.com/trial-closes-

tie-downs-and-gaining-agreements.

- Yale, Aly J. "Are Real Estate Agents Still Relevant In The Age Of Tech?" *Forbes*, Forbes Magazine, 1 Aug. 2018, www.forbes.com/ sites/alyyale/2018/08/01/are-real-estate-agents-still-relevant-in-the-age-of-tech.

EPILOGUE - WHAT HAPPENS NEXT?

- "Benjamin Franklin Quotes." *BrainyQuote*, Xplore, www. brainyquote.com/quotes/benjamin_franklin_383997.
- Krippendorff, Kaihan. "7 Strategic Openings To Innovate, Grow & Dominate (Part 1)." *Growth Institute*, blog.growthinstitute. com/business-strategy/7-strategic-openings-part-1.
- Vaynerchuk, Gary. "Jab, Jab, Jab, Right Hook: How to Tell Your Story in a Noisy Social World." *Amazon*, Amazon, 26 Nov. 2013, www.amazon.com/Jab-Right-Hook-Story-Social/ dp/006227306X.
- "Walt Disney Quotes." *BrainyQuote*, Xplore, www.brainyquote. com/quotes/walt_disney_130027.

INDEX

U

V

W

INDEX

ABOUT THE AUTHOR

James **Tyler**, founder, and CEO of Marketing Engines, Inc. is a leading marketing expert for the mortgage and the real estate industries. He has led many financial organizations as the Vice President of Sales and the Director of Marketing. His marketing, sales, mortgage, and real estate knowledge have given him significant leverage to help financial businesses find traction and growth.

He is a sought out author and speaker who presents at conferences for startups, lenders, mortgage companies, and real estate firms. His business strategies and coaching programs have helped hundreds of financial individuals and real estate professionals nationwide find wealth and happiness.

James, his wife, Randa, and their daughter, Liana live in Orange County, California.

For book inquiries, visit Dominate.RealEstate.
For Business Consulting & Strategic Coaching, visit JamesTyler.com.
For Web Development & Digital Marketing, visit MarketingEngines.com.